Characterization of Ceramic-Ferrite Magneto-Electric Composites

Dr. R. Saravanan, M.Sc., M.Phil., Ph.D.

Associate Professor
Research Centre and PG Department of Physics
The Madura College (Autonomous)
Madurai – 625 011

Published by **Materials Research Forum LLC**
Millersville, PA 17551, USA

Published as part of the book series
Materials Research Foundations
Volume 136 (2023)
ISSN 2471-8890 (Print)
ISSN 2471-8904 (Online)

Print ISBN 978-1-64490-218-9
ePDF ISBN 978-1-64490-219-6

Distributed worldwide by

Materials Research Forum LLC
105 Springdale Lane
Millersville, PA 17551
USA
http://www.mrforum.com

Printed in the United States of America
10 9 8 7 6 5 4 3 2 1

Table of Contents

Preface

Magneto-electric composites are materials with magneto-electric property which is a common property present between the different materials making the composite. Different materials with different properties are combined in a particular ratio to form the composite. In magneto-electric composite, the ferrite with magnetic property and the ferroelectric with the electric property are combined to produce the magneto-electric property. The ME composites are very important materials for designing new microwave sensors, e.g., field probes, and devices, e.g., filters, attenuators, capacitive resonators and gyrators. These materials are also used to design new devices in the field of medicine.

These samples are characterized by powder X-ray diffraction for the determination of the structural details of the synthesized composites. Further, the powder XRD experimental data are utilized in the profile refinement techniques for which standard software programs are being used. The results are used for the electron density analysis of the grown composites and the changes in the electron density distributions for composition x. The SEM results are used for the determination of particle sizes of the synthesized composites and the XRD analysis of the grain sizes will be compared with the particles sizes obtained from SEM analysis. The optical band gaps of the prepared composites are studied using UV-Vis spectroscopy and the variation of the bandgap for different composites and composition x, are also studied. The electrical characterizations of the composites are done by the frequency vs dielectric constant, dielectric loss, and the P-E characterizations. The change in the magnetic property with the composite type (nature of ferrite addition in the composite) and concentration x, are studied using VSM (Vibrating Sample Magnetometer).

Chapter I reports about the nature of the materials used for preparing the magnetoelectric composite. The characterization techniques used to investigate the properties of the prepared materials to analyze the properties. The instruments used in the different characterization techniques. The sample preparation method for all the composites prepared. The research methodologies used to study the structure, characterization, and charge densities between the atoms of the composite.

Chapter II depicts the results of the four series of magnetoelectric composite, prepared by the solid-state reaction process. (1-x) $BaTiO_3$ + x MFe_2O_4 (M = Co, Zn, Ni, Mg and x = 0.2, 0.4, 0.6, 0.8). The raw profiles for various compositions in a series of the XRD data are presented. The refined profiles of the XRD data (Rietveld fitted profiles), SEM micrographs, EDS spectra, three-dimensional (3D), two-dimensional (2D), and one-dimensional (1D) electron density distribution from the (Maximum Entropy Method)

MEM method are presented in this chapter. UV-Vis absorption graphs Tauc plots to find the optical bandwidth. The electrical characterization: Capacitance vs. frequency graph, P-E hysteresis graph, and the magnetic characterization; M-H hysteresis plot.

Chapter III gives a detailed analysis of all the characterizations observed for the prepared composites. The details of sample preparation are given in sec 3.2. The structural analysis of the prepared composites using powder XRD data is given in sec3.3. Section 3.4 explains the morphological nature of the composites from the SEM images. Also gives the details of particle and grain sizes of the prepared composites. The elementary compositions of the composites have been discussed in sec 3.5. The optical characterization from the UV-Vis spectral data is given in sec 3.6. The electrical characterization; dielectric, dielectric loss, and P-E hysteresis are discussed in sec 3.7. Sec 3.8 gives the magnetic characterization of the composites from M-H hysteresis data. The electron density studies depicting the electronic structure, nature of bonding, electron densities correlation with the observed properties are explained in 3.9.

The analysis of the observed results, correlation with the electron density studies are discussed in detail

Chapter IV gives the major conclusion from all the observed characterizations; structural analysis from powder XRD, Morphological and grain size from the SEM, elementary analysis from EDS, optical band gap from UV-Vis spectral analysis, the electrical nature of the sample from dielectric, dielectric loss and P-E characterization and the magnetic characterization from the M-H hysteresis.

The role of electron densities in the observed properties is also given in this chapter.

Results of the work have been published as follows:

Charge density analysis, Structural, Electrical and Magnetic studies of $(1-x)BaTiO_3+xNiFe_2O_4$ ceramic composite, S.V. Meenakshi, **R. Saravanan**, N. Srinivasan, D. Dhayanithi, Nambi Venkatesan Giridharan, *Journal of Electronic materials*, Springer Publications; Vol No. 49, 7349-7362 (2020) (IF 1.938)

Exploration of $(1-x)BaTiO3 + xZnFe2O4$ magneto-electric ceramic composite on charge density: Structure and its characterization, S.V. Meenakshi, **R. Saravanan**, N. Srinivasan, D. Dhayanithi , Nambi Venkatesan Giridharan, *Journal of Alloys and Compounds* (Elsevier Publications) Vol. No. 888, 161491 (2021)(IF 5.316)

List of papers presented in conferences

Study of the magnetic behavior of $MgFe_2O_4$ and its composites, T. Suganya, A. Vinothini, S.V. Meenakshi, **R. Saravanan**

UGC sponsored one day International Conference on "Recent trends in Materials Science and Applications" organized by the Department of Physics, Sri Meenakshi Government Arts College for Women, Madurai 625 002, on January 6th, 2017, ISBN: 819331402-6.

Chapter 1

Introduction

Abstract

In this book, some magneto-electric composites based on $BaTiO_3$ ceramics have been synthesized and reported. Four series of samples have been synthesized as follows:

1. $(1-x) BaTiO_3 + x NiFe_2O_4$, 2. $(1-x) BaTiO_3 + x ZnFe_2O_4$, 3. $(1-x) BaTiO_3 + x CoFe_2O_4$, 4. $(1-x) BaTiO_3 + x MgFe_2O_4$

These samples are characterized by powder X-ray diffraction for the determination of the structural details of the synthesized composites from the Rietveld refinement and JANA 2006. Further, the powder XRD experimental data are utilized in the profile refinement techniques with standard software programs (MEM) by using the software PRIMA and VESTA. The results are used for the electron density analysis of the grown composites and the changes in the electron density distributions with respect to composition x. The SEM results are used for the determination of particle sizes of the synthesized composites and the XRD analysis of the grain sizes using GRAIN software, are compared with the particles sizes obtained from SEM analysis.

The optical band gap (E_g) of the prepared composites is studied using UV-Vis spectroscopy and the variation of the bandgap with respect to the composite type and concentration x, are also analyzed. The electrical characterizations of the composites are done by the frequency vs. dielectric constant, dielectric loss, and the P-E characterizations. The magnetic behavior and the change in the magnetic behavior with respect to the composite type (nature of ferrite addition in the composite) and concentration x, are studied using VSM (Vibrating Sample Magnetometer).

Magneto-electric materials are composite materials since they contain two different materials with different characteristics (Ferroelectric and Ferromagnetic materials). The two materials have separate characteristics either ferroelectric or ferrite but will be present as one material and hence are known as composite materials. Ferroelectric materials are materials with electrical characteristics (dielectric, piezoelectric, p-E characteristics) and ferromagnetic materials are also known as ferrites with the magnetic property. Since materials with magnetic properties are used for designing storage devices and ferroelectric materials have numerous applications and various properties of ferroelectric, piezoelectric, and dielectric.

Keywords

Magneto electric, $BaTiO_3$, $NiFe_2O_4$, $ZnFe_2O_4$, $CoFe_2O_4$, $MgFe_2O_4$

1.1 Objectives

The objective of the present work is to synthesize and characterize the magneto-electric composites based on $BaTiO_3$ ceramics and to correlate the samples that are synthesized with the electron density studies. To achieve the aims; the following series have been synthesized by solid-state synthesis technique.

1. $(1-x) BaTiO_3 + x NiFe_2O_4$

2. $(1-x) BaTiO_3 + x ZnFe_2O_4$

3. $(1-x) BaTiO_3 + x CoFe_2O_4$

4. $(1-x) BaTiO_3 + x MgFe_2O_4$

These samples are characterized by powder X-ray diffraction for the determination of the structural details of the synthesized composites from the Rietveld refinement (Rietveld 1969) and JANA 2006 [Petricek et al., 2014]. Further, the powder XRD experimental data are utilized in the profile refinement techniques with standard software programs (MEM) [Collins, 1982] by using the software PRIMA [Izumi, 2002] and VESTA [Momma, 2008]. The results are used for the electron density analysis of the grown composites and the changes in the electron density distributions with respect to composition x. The SEM results are used for the determination of particle sizes of the synthesized composites and the XRD analysis of the grain sizes using GRAIN software [Saravanan, Personal communication] are compared with the particles sizes obtained from SEM analysis.

The optical band gap (E_g) of the prepared composites are studied using UV-Vis spectroscopy and the variation of the bandgap with respect to the composite type and concentration x, are also analyzed.

The electrical characterizations of the composites are done by the frequency vs. dielectric constant, dielectric loss, and the P-E characterizations.

The magnetic behavior and the change in the magnetic behavior with respect to the composite type (nature of ferrite addition in the composite) and concentration x, are studied using VSM (Vibrating Sample Magnetometer).

1.2 Composites

Two or more materials with different characteristics make the composite material. Such materials are combined to form a material with properties different from individual materials. The individual materials remain separate and distinct, in a combined structure. This makes a difference between the composites from mixtures and solid solutions [Fazeli et al., 2018 Elhajjar et al., 2017].

Recent research work is on producing good sensing materials, computation, and communication materials from composites [McEvoy et al., 2015]. Composite materials are used as building materials for buildings, bridges, car bodies, etc. The composite materials are used in general automotive applications. The most advanced examples perform routinely on spacecraft and aircraft to overcome the effects of environments on the aircraft [Fazeli et al., 2018].

Ceramic composites are mixing of ceramic materials, which will give a combined effect as a new property of the material. When a ferroelectric and a ferrite material are combined, they give a new effect known as the Magnetoelectric effect. Hence such materials are magneto-electric ceramic composites.

1.3 Magnetoelectric ceramic composites

Magnetoelectric composites are materials in which Magneto-electric property (ME) is observed [Van den Boomgaard et al., 1976]. High magneto-electric coefficient and good coupling between ferroelectric phase and ferrite phase are the important reasons to attract the investigators than single-phase materials [Narendrababu et al., 2005]. The presence of two phases such as ferrimagnetic and ferroelectric composites has magnetoelectricity or ME effect when there is the magnetization of the material by an electric field or polarization by a magnetic field. This magnetoelectric effect takes place in the composites and is absent in an individual constituent phase [Mitoseriu et al., 2007]. H. Schmid in 1994 [Schmid, 1994] was the first researcher to use the term multiferroic. This term multiferroic represents materials that exhibit at least two ferroic properties (ferroelectricity, ferromagnetism, and ferroelasticity) at the same time. The Magneto-electric response is given by the equations: $\Delta P = \alpha\ \Delta H$ or $\Delta E = \alpha E \Delta H$, where E denotes electric field and α is the effective ME coefficient. Magneto dielectric effect is the change in the dielectric constant with the magnetic field and the capacitive nature of the materials can be investigated. Multiferroic materials are important and interesting due to their strong applications as sensors, transducers, filters, oscillators, phase shifters, memory devices, etc. [Neaton et al., 2005 Cheong et al., 2007].

Magneto-electric materials are composite materials since they contain two different materials with different characteristics (Ferroelectric and Ferromagnetic materials). The two materials have separate characteristics either ferroelectric or ferrite but will be present as one material and hence are known as composite materials. Ferroelectric materials are materials with electrical characteristics (dielectric, piezoelectric, p-E characteristics) and ferromagnetic materials are also known as ferrites with the magnetic property. Since materials with magnetic properties are used for designing storage devices and ferroelectric materials have numerous applications and various properties of ferroelectric, piezoelectric, and dielectric. It becomes an option to be used in applications such as multilayer capacitors and electro-optic devices.

1.3.1   Ferrites

The general formula for ferrites is XY_2Z_4, where X occupies the tetrahedral sites and Y the octahedral sites. X^{2+} is the divalent positive metal ion like Fe, Co, Zn, Ni, Mg, Mn, etc., and Y^{3+} is a trivalent cation as In, Al, Ti, V, Cr, Co, Fe, Ni, etc. and Z^{2-} is the divalent oxygen ion. Memory storage devices, filters circuit, computer machinery, transformer cores, electronic circuits power delivering devices, cancer disease treatment, and magnetic resonance imaging [Gul et al., 2012] are some of the applications of ferrites in many fields. As an important member of the ferrite family, zinc ferrite ($ZnFe_2O_4$) is special which attracts significant research interest due to its structural and magnetic properties. $ZnFe_2O_4$ is a chemically and thermally stable semiconductor material [Yang et al., 2014]. Bulk $ZnFe_2O_4$ has a normal spinel structure with the tetrahedral (X^{2+}) sites occupied by Zn^{2+} and the octahedral (Y^{3+}) sites occupied by Fe3+ [Masoundpanah et al., 2014]. In the "inverse" spinel structure, X^{2+} cations are replaced by half of the Fe^{3+} ions in the tetrahedral sites and the second half of the Fe^{3+} ions remaining at the octahedral sites [Seyyed Ebrahimi et al., 2014 Ben Ali et al., 2016].

The spinel ferrite having the formula AB_2X_4 is one of the most interesting and important families of crystalline compounds, as the color, diffusivity, magnetic behavior, conductivity, and catalytic activity are observed from the occupation of metals at octahedral and tetrahedral sites [Hou et al., 2010].

Cobalt ferrite (CFO) is an attractive material for its high coercivity, moderate saturation magnetization, and large anisotropy. Brabers have found that $CoFe_2O_4$ (CFO) has the largest magnetostriction among magnetic materials that do not contain any rare-earth elements [Brabers, 1995].

1.3.2   Ferroelectric materials

Ferroelectric materials such as; barium titanate, lead titanate, lead zirconate titanate, have perovskite-type structures. The applications of ferroelectric ceramics are in the areas of dielectric ceramics for capacitor applications, especially ferroelectric thin-film technology [Babu Naidu et al., 2015]. The structure of the perovskite family is of the type ABO_3 [Candra Babu Naidu et al., 2015]. The ferroelectric behavior is investigated with the hysteresis curve from which the extent of polarization is measured with the applied electric field.

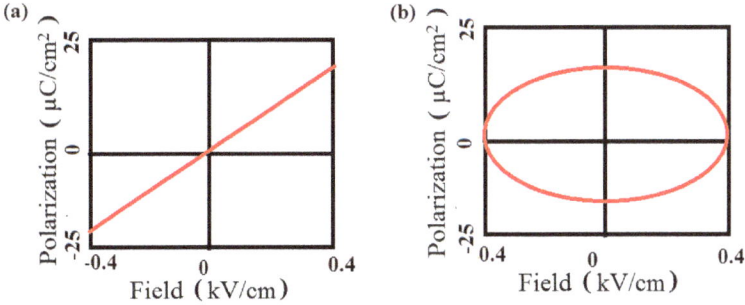

Figure 1.1 *P-E hysteresis response (a) Ideal linear capacitor (b) Ideal resistor response*

Figure 1.1 represents the response of the ideal capacitor and a resistor. For the ideal capacitor response, the polarization increases linearly with the increase in the applied electric field. The polarization decreases with the decrease in the applied field and reaches a minimum and then increases when the electric field decreases. When the field is reversed, the polarization also gets reversed.

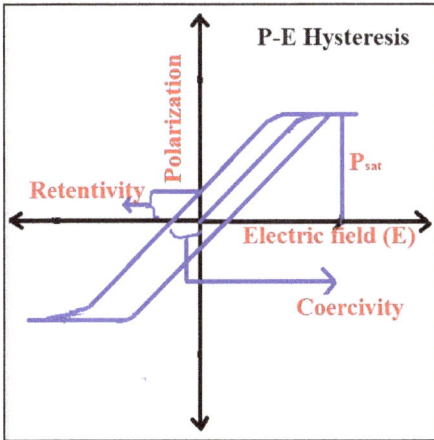

Figure 1.2 *P-E hysteresis curves for a ferroelectric material.*

Figure 1.2 shows the response of a ferroelectric material. The polarization of the material increases with the increase in the strength of the applied electric field and reaches a maximum and remains constant. This maximum value which becomes a constant is saturation polarization. When the field is decreased, the polarization also decreases and has a value for the null electric field and this value is retentivity. For the increase in the reverse field, the polarization reaches a zero value, and this field for which the polarization becomes zero is called coercivity.

1.4 Importance of magnetoelectric ceramic composites

Magnetoelectric composites and the devices from them have been an important topic of research because of many applications from low power sensing to high power converters. The development of ME-based energy harvesters and ME gyrators for power applications including power efficiency, power density, and figures of merit are important applications of magnetoelectric materials. ME gyrator satisfies the needs of power conversion with the highest efficiency (> 90%) for potential uses in power applications [Chung Ming Leung et al., 2018] ME composites are used in different technological applications such as magnetic sensors [Wang et al., 2011], micro-electro-mechanical systems (MEMS), and energy harvesters [Marauska et al., 2018 Dong et al., 2008]. The ME effect is very important for designing new microwave sensors, field probes, and devices, filters, attenuators, capacitive resonators, gyrators. The experimental realization of a composite ME material with enhanced microwave properties remains a central pursuit in materials [Castel et al.] Modern medicine practices various passive (for example, bone replacement) and active implants (apparatuses for auxiliary circulation, artificial heart, various stimulants, etc.). Non-invasive testing of functional characteristics, resources, and other properties are possible using high-sensitivity magnetic field sensors.

1.5 Structure of BaTiO$_3$

Barium titanate is a ferroelectric ceramic material, with piezoelectric, dielectric, photorefractive properties.

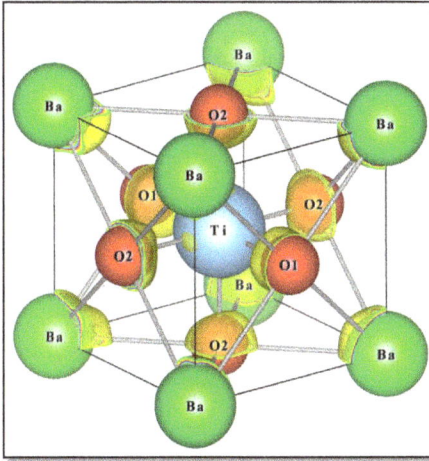

Figure 1.3 *Structure of tetragonal BaTiO₃ (Barium titanate)*

The solid can exist in five phases, from high-temperature to low temperature: hexagonal, cubic, tetragonal, orthorhombic, and rhombohedra crystal structure. The cubic phase does not exhibit ferroelectric property whereas all of the other phases exhibit the ferroelectric effect. The high-temperature $BaTiO_3$ assumes a cubic phase, with Ti vertices and Ti-O-Ti edges. Ba^{2+} is at the center with a nominal coordination number of 12 in the cubic phase. At lower temperatures, $BaTiO_3$ takes the other phases, associated with the movement of the Ba^{2+} to the off-center position. Figure 1.3 shows the structure of $BaTiO_3$; tetragonal phase.

1.5.1   Properties and applications

Barium titanate is an electrical insulator in its purest form, can act as a dielectric ceramic material for capacitors. The Thermistor is one of the applications of barium titanate. Being a piezoelectric material is used in microphones and other transducers. Thermal cameras use some varieties of un-cooled sensors manufactured from the pyroelectric properties. Photorefractive applications use barium titanate. Thin films of barium titanate show electro-optic modulating at 40Ghz.

1.6 Structure of ferrites (XFe_2O_4; X = Ni, Zn, Co, Mg)

Ferrites are formed with high proportions of iron(III) oxide blended with small proportions of metallic elements, like strontium, barium, manganese, nickel, and zinc, ceramic ferrite material is produced [Carter et al., 2007]. Most of the ferrites are not electrically conductive but are useful in applications like magnetic cores for transformers to suppress eddy currents [Spaldin et al., 2010]. Most of the ferrites with spinel structure have the formula AB_2O_4, where A and B represent various metal cations, including iron (Fe). Spinel ferrites usually adopt a crystal structure consisting of cubic close-packed (fcc) oxides (O_2-) with A cations occupying one-eighth of the tetrahedral holes and B cations occupying half of the octahedral holes, i.e., $A^{2+}B^{3+}_2O^{2-}_4$. Ferrite crystals do not assume the ordinary spinel structure, but the inverse spinel structure: B cations occupy one-eighth of the tetrahedral holes, A cations occupy one-fourth of the octahedral sites, the other one-fourth by B cation. The magnetic material $ZnFe_2O_4$, an example of structure spinel ferrite has Fe^{3+} at the octahedral sites and Zn^{2+} occupying the tetrahedral sites [Shriver et al., 2006]. Some ferrites like barium and strontium ferrites assume hexagonal structures [Ullah et al.].

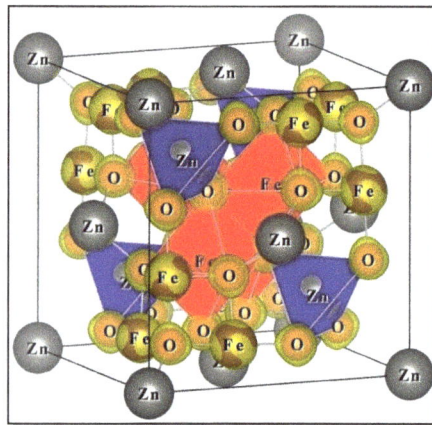

Figure 1.4 *Structure of $ZnFe_2O_4$ a ceramic ferrite: shows the tetrahedral and octahedral sites occupied by Zn and Fe.*

1.6.1 Properties and applications

Hard ferrites have a high coercivity and are difficult to demagnetize. **Strontium ferrite**, $SrFe_{12}O_{19}$ ($SrO \cdot 6Fe_2O_3$), used in small electric motors, micro-wave devices, recording media, magneto-optic media, telecommunication, and electronic industry [Ullah et al., 2013]. Strontium hexaferrite ($SrFe_{12}O_{19}$) has high coercivity which is the property of a hard magnet. The hard ferrites are mostly used in industrial applications as permanent magnets and micro and nano-types systems such as biomarkers, bio diagnostics, and biosensors [Gubin et al, 2005]. Barium ferrite ($BaFe_{12}O_{19}$), is also a permanent magnetic material having the property of a hard ferrite.

Semi-hard ferrites; Cobalt ferrite, $CoFe_2O_4$, is in between soft and hard magnetic material and is a semi-hard material [Hosni, 2016]. It is mainly used for its magnetostrictive applications like sensors and actuators [Olabi, 2008]. It has high saturation magnetostriction (~200 ppm). $CoFe_2O_4$ has also the benefits to be rare-earth-free, which makes it a good substitute for Terfenol-D [Sato Tortelli et al., 2014]. Its magnetostrictive properties can be tuned by inducing a magnetic uniaxial anisotropy [Slanczewiski, 1958]. The magnetoelectric effect in cobalt ferrite is enhanced from its induced magnetic anisotropy [Aubert, 2017].

Soft ferrites have low coercivity, so their magnetization can be easily changed and become conductors of magnetic fields. High-frequency inductors, transformers, antennas, and various microwave components, are some of the applications of soft ferrites. MnZn has higher permeability and saturation induction than NiZn. Figure 1.4 shows the structure of $ZnFe_2O_4$ ferrite.

1.7 Solid-state reaction technique

The raw chemicals in the powder form, are taken in the stoichiometric ratio. The weighed powders are mixed and made into pellets. Solids do not react together at room temperature over time scales, and it is necessary to heat them to higher temperatures, often to 1000 to 1500 °C, for the reaction to occur at an appreciable rate. The factors on which the solid-state method depends on the structural properties of the reactants, the surface area of the solids, their reactivity, and the thermodynamic free energy change associated with the reaction.

1.7.1 Mixing/Grinding

The weighed reactants are mixed by grinding the samples. For the mixing of small quantities, an agate mortar and pestle are used. Organic liquid acetone or alcohol is added to the mixture forming a paste when mixed thoroughly and during grinding, the organic liquid gradually evaporates after 10 to 15 minutes. A ball mill is used for mechanical mixing for quantities higher than 20g, and the process may take several hours. The solid-state reaction starts, while mixing/grinding since the size of the particles decreases and the change in energy is more when the radius of curvature is a few micrometers. Figure 1.5 shows the grinding tool agate mortar and pestle.

Figure 1.5 *Grinding tool; Agate mortar & Pestle.*

1.7.2   Pressing/Pelletizing

Pelletizing of samples is done before heating since it increases the area of contact between the grains. During pelletizing, the solid particles come closer and the solid-solid interaction starts. The powder samples can be pressed into dense pellets by applying pressure using a hydraulic press. The pressure applied depends upon the composition and nature of the samples.

1.7.3   Heat treatment (Sintering)

The heat treatment to be carried out depends very much on the reactivity of the reactants. By proper control of the temperature or the atmosphere, the nature of the reactant chemicals is considered for sintering. Depending on the nature of the materials to be sintered the temperature or the heating atmosphere is decided. The melting points of the sintering materials are to be considered to fix the temperature of sintering. Depending on the reacting capacities of the samples the duration of sintering is decided. A high efficient furnace is used for heat treatment. Alumina combustion boat inert to hydrogen, carbon, and refractory metals can be used at operating temperatures to $1750^{\circ}C$ in the oxidizing and reducing atmospheres.

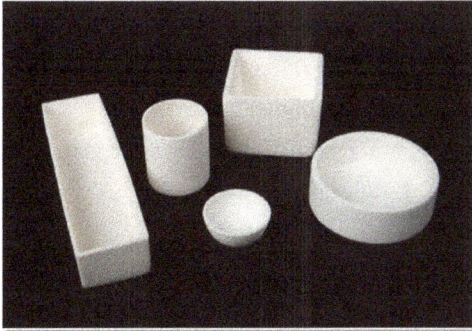

Figure 1.6 *Alumna boats for sintering of the samples.*

Alumna boats for sintering the samples are shown in Figure 1.6. Figure 1.7 shows the picture of a tubular furnace at Madura College used to sinter our samples around 1350 °C. These boats were purchased from VB Ceramics, Chennai, Tamilnadu; India. Since the samples were to be sintered around 1350°C, alumna boats were used as the sample container for the high-temperature sintering of the samples.

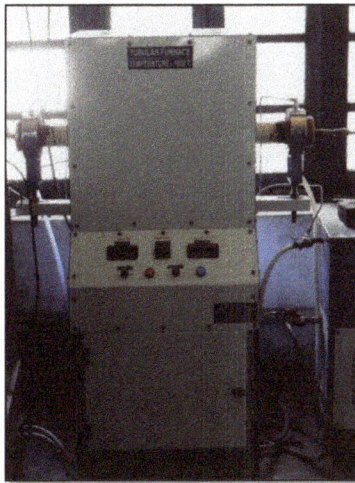

Figure 1.7 *tubular furnace for sintering up to 1600° C*

1.8   Preparation of Magneto-electric ceramic composites.

Magneto-electric ceramic composites are prepared by synthesizing ferrites and ferroelectrics separately. To prepare a ferrite or a ferroelectric material, the reactants are taken in the stoichiometric ratio, mixed, and sintered to form the product. In this work, the ferroelectric material is Barium titanate ($BaTiO_3$) and the ferrites are XFe_2O_4 where X = Ni, Co, Zn, and Mg.

1.8.1   Preparation of $(1-x)BaTiO_3 + xNiFe_2O_4$

The solid-state reaction method is used to synthesize the ceramic composite of $NiFe_2O_4$ and $BaTiO_3$. The ferroelectric phase ($BaTiO_3$) was synthesized by taking suitable quantities of $BaCO_3$ (Alfa Aesar99.99%) and TiO_2 (Alfa Aesar 99.99%) in stoichiometric proportion and mixing them thoroughly using an agate mortar for 4 h. The mixture was then pelletized and sintered for 2 h at 1350°C in the air. NiFe2O4 powder (99.99%) purchased from Alfa Aesar was used as a ferrite phase. The ferroelectric-ferrite composite samples were prepared by mixing nickel ferrite and barium titanate with proportions (where x = 0.20, 0.40, 0.60, 0.80) according to the formula $(1 - x) BaTiO_3 + x NiFe_2O_4$. The powders of barium titanate and nickel ferrite were ground using an agate mortar for 12 hours. After mixing for 12 hours the mixed powder was then pressed to form pellets with 10 mm and 2 mm uniform diameter using a pelletizing machine. These pellets were sintered at 1250°C for 5 h in the air at the rate of 5°C/min.

1.8.2   Preparation of $(1-x)BaTiO_3 + xZnFe_2O_4$

The ceramic composite of $ZnFe_2O_4$ and $BaTiO_3$ was prepared by the solid-state reaction method. Barium titanate ($BaTiO_3$) was synthesized by mixing $BaCO_3$ (Alfa Aesar 99.99%) and TiO_2 (Alfa Aesar 99.99%) in stoichiometric proportion and thoroughly using agate mortar for 4 hours. The mixture was then pelletized and sintered for 2 hours at 1350 °C in air to synthesize $BaTiO_3$. $ZnFe_2O_4$ powder (99.99%) purchased from Alfa Aesar was used as a ferrite phase. The samples of the magnetoelectric composite were prepared by mixing $ZnFe_2O_4$ and $BaTiO_3$ according to the formula $(1-x) BaTiO_3 + x ZnFe_2O_4$ with proportions (where $x = 0.20, 0.40, 0.60, 0.80$). The barium titanate and zinc ferrite were mixed using agate mortar for 12 hours and then pressed to form pellets with 10 mm and 2 mm uniform diameter and thickness, using a steel die and hydraulic press. These pellets were finally sintered at 1200 °C for 5 hours in the air at the rate of 5 °C/min to synthesize the composite $(1-x) BaTiO_3 + x ZnFe_2O_4$. Finally, sintered products are sent for characteristic studies after grinding them into powders.

1.8.3   Preparation of $(1-x)BaTiO_3 + xCoFe_2O_4$

$(1-x)BaTiO_3 + xCoFe_2O_4$ ceramic composite was prepared by solid-state reaction method. Stoichiometric quantities of CoO (Alfa Aeser 99.99%) and Fe_2O_3 (Alfa Aeser 99.99%) were weighed and mixed thoroughly using an agate mortar for 12 hours to prepare the ferrite phase. The well-mixed powder of CoO and Fe_2O_3 was then made into pellets and sintered at 1100 °C for 12 hours in the air atmosphere. To synthesize the ferroelectric phase; suitable quantities of $BaCO_3$ (Alfa Aesar 99.99%) and TiO_2 (Alfa Aeser 99.99%) are taken in stoichiometric proportion. These powders were weighed and mixed thoroughly using agate mortar for 4 hours. The mixture was pelletized and sintered for 2 hours at 1350 °C in the air atmosphere. By mixing cobalt ferrite and barium titanate in proportions according to the formula, the ceramic composite of $(1-x)BaTiO_3 + xCoFe_2O_4$ for x = 0.2, 0.4, 0.6, 0.8 were prepared. The barium titanate and cobalt ferrite were mixed using agate mortar for 12 hours and pressed in the form of pellets of 10 mm diameter and 2 mm thickness, using a steel die and hydraulic press and sintered for 12 hours at 1150 °C at a rate of 5 °C/min.

1.8.4   Preparation of $(1-x)BaTiO_3 + xMgFe_2O_4$

The composite material was prepared by the solid-state reaction method. Barium titanate ($BaTiO_3$) was prepared by mixing $BaCO_3$ (Alfa Aesar 99.99%) and TiO_2 (Alfa Aesar 99.99%) in stoichiometric proportion and thoroughly using agate mortar for 4 hours. The mixture was then pelletized and sintered for 2 hours at 1350 °C in air. Similarly, to synthesize $MgFe_2O_4$, MgO (Alfa Aesar 99.99%) and Fe_2O_3 (Alfa Aesar 99.99%) powders were mixed in stoichiometric ratio for 12 hours using agate mortar. The mixture was pelletized and sintered at 1200 °C for 5 hours in the air at the rate of 5 °C/min.

The samples of the magnetoelectric composite were prepared by taking MFO and BTO with proportions (where x = 0.2, 0.4) in accordance with the formula $(1-x)BaTiO_3 + xMgFe_2O_4$ and were mixed using agate mortar for 12 hours and then pressed to form pellets with 10 mm and 2 mm uniform diameter and thickness, using a steel die and hydraulic press. These pellets were sintered at 1100 °C for 15 hours in the air at the rate of 5 °C/min.

1.9   Characterization methods and the instruments used

X-ray diffraction, SEM-EDS, UV-Vis, Variation of capacitance with frequency at room temperature, and P-E hysteresis (Electrical characterization), M-H hysteresis studies (Magnetic characterization) are the characterizations used in this research work.

1.9.1   Powder X-ray diffraction

Crystalline solids are made of regular arrays of atoms, ions, or molecules with the interatomic spacing on the order of 100 pm or 1 Å. The incident X-ray should have a wavelength compared to the interatomic spacing of the atoms in the crystals or crystallites.

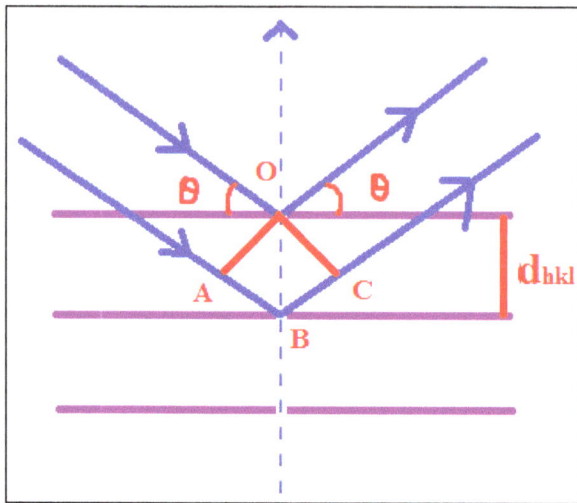

Figure 1.8 *X-ray diffraction shows the path difference between the two rays.*

From figure 1.8, the path length difference = AB + BC; AB = BC = $d_{hkl} \sin\theta$; Must be an integral number of wavelengths, $n\lambda = 2d_{hkl}\sin\theta$ (n = 1, 2, 3, …) is the Bragg Equation, $\lambda = 2d_{hkl}\sin\theta$ for n=1. A powder is composed of many small ground crystals, known as crystallites, are randomly oriented to one another is placed in the path of a monochromatic X-ray beam incident on the powder will get diffracted from the planes of the crystallites that are oriented at the angle to fulfilling the Bragg condition. The angle between the diffracted beams and the incident beam is 2θ.

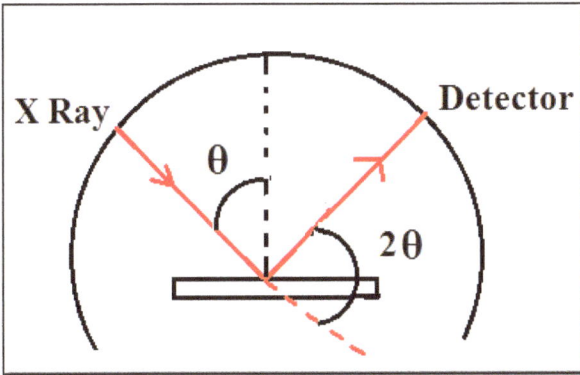

Figure 1.9 *Schematic diagram of X-ray scattering.*

Figure 1.10 *X-ray diffractometer: Bruker AXS D8*

The XRD data for all the samples are taken from the SAIF-STIC, Cochin, Kerala; India. The XRD instrument used at the SAIF is Bruker AXS D8. The 2θ readings are observed from 0° to 120° for all the samples.

1.9.2   UV-Visible Spectroscopy

Ultraviolet-visible spectroscopy (UV–Vis) is the absorption spectroscopy or reflectance spectroscopy in the ultraviolet region, adjacent visible regions of the electromagnetic spectrum. Atoms and molecules undergo electronic transitions either through absorption or emission where there will be a transition of atoms from the ground state to excited or excited to the ground state. Hence absorption spectroscopy is complementary to fluorescence spectroscopy [Skoog, Douglas, et al., 2007].

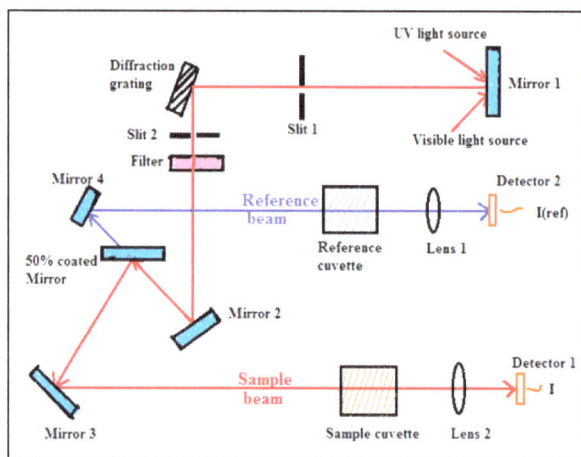

Figure1.11 *Block diagram of UV-Vis spectrometer.*

The optical bandgap is found using UV-Vis absorption spectroscopy. From the bandgap, the physical characteristics can be studied like conductivity, nature of the material, the strength of the material, porosity, electrical nature of the material. The thickness and optical properties of thin films are investigated from UV–Vis spectroscopy in the semiconductor industry. Absorption of the ultra-violet radiations excites the electrons from the ground state to a higher energy state. Figure 1.11 shows the block diagram of the UV-Visible spectrometer which explains the working of a UV-Visible spectrometer. The energy of the ultraviolet radiation that is absorbed is equal to the energy difference between the ground state and higher energy states   ($\Delta E = h\gamma$). Figure 1.12 shows the UV-Vis instrument Varian Cary 5000 at SAIF, STIC, Cochin, Kerala; India.

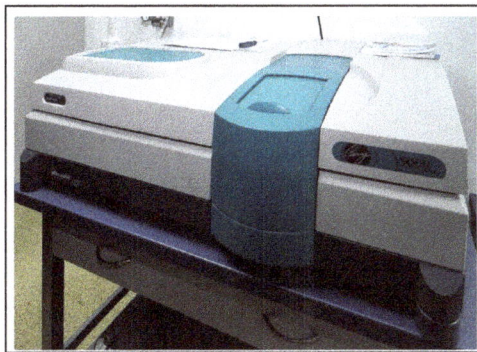

Figure 1.12 *UV-Visible spectrometer; Varian Cary 5000*

1.9.3   Scanning Electron Microscope (SEM)

A scanning electron microscope (SEM) is used to study the morphology of the sample because it focuses electrons on the samples and collects the secondary electrons which are emitted from the surface of the sample. The focused electrons interact with atoms of the sample. The various signals produced from the interactions of the electrons with the atoms contain information about the surface topography and composition of the sample. Researchers have found that in a conventional SEM the specimens are observed in high vacuum, or in environmental SEM the specimens are observed in low vacuum or wet conditions in a variable pressure [Stokes Debbie, 2008]. Though the secondary electrons (SE) with the other types of electrons are emitted, the secondary electrons are only collected by the common SEM instrument. All SEM instruments have secondary electron detectors.

SEM samples are specially prepared to fit specimen space and also that they can withstand the high vacuum conditions and the high energy beam of electrons. Samples are mounted rigidly on a specimen holder using a conductive adhesive. The secondary electrons are emitted from the surface of the sample and hence SEM is used to study the morphology of the sample [Goldstein et al., 1981]. The electrons are detected by an Everhart-Thornley detector, [Everhart et al., 1960]. The detector has a collector, scintillator, and photomultiplier system. The process of obtaining SEM is shown as the schematic diagram in Figure 1.13 and the SEM instrument JOEL JSM- 6390 LV/LGS in Figure 1.14 at SAIF, STIC, Cochin, Kerala; India. The electrons from the electron source are accelerated using electric condensing lenses, scan coils, and are focused on the specimen or the sample. The low energy secondary electrons are emitted and are collected by the collector of the detector, then accelerated towards the scintillator through a positive voltage. The electrons have acquired sufficient energy through accelerations and produce scintillations which are sent to the photomultiplier tube for amplification. The amplified signals are recorded for the image.

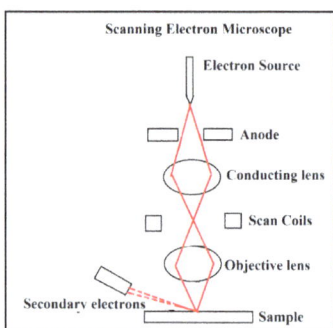

Figure 1.13 Schematic diagram of SEM instrument.

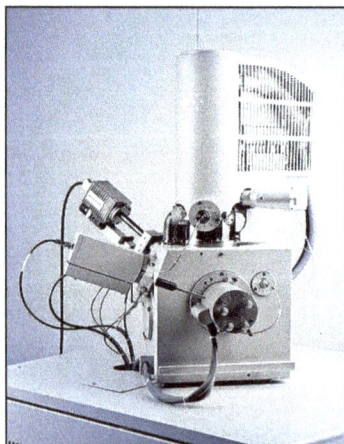

Figure 1.14 SEM-EDS instrument; JOEL JSM- 6390 LV/LGS

1.9.4 Energy-dispersive X-ray Spectroscopy

Energy-dispersive X-ray spectroscopy (EDXS) is a technique to find the elemental composition present in material for elemental analysis and chemical characterization of a sample. The interaction of the source of X-ray and a sample gives elements present in the sample. The atomic structure that has a unique set of peaks on its electromagnetic emission spectrum is unique for each element. Moseley's law is used to predict the peak positions with more accuracy better than the experimental resolution of a typical EDXS instrument. The incident beam focused into the sample excites an electron in an inner shell, ejecting it from the shell. The ejected electron creates an electron-hole and the electron from a higher-energy shell fills the hole. The difference in energy between the higher and the lower energy is emitted in the form of an X-ray. EDXS measures the number and energy of the X-rays emitted from a specimen. The energies of the X-rays depend on the difference in energy between the two shells and of the atomic structure of the emitting element, hence EDS allows the elemental composition of the specimen to be measured. When the excess energy is transferred to a third electron from an outer shell it gets ejected and is called an Auger electron. The method of observing the spectrum is known as Auger electron spectroscopy (AES) [Jenkins et al., 1982].

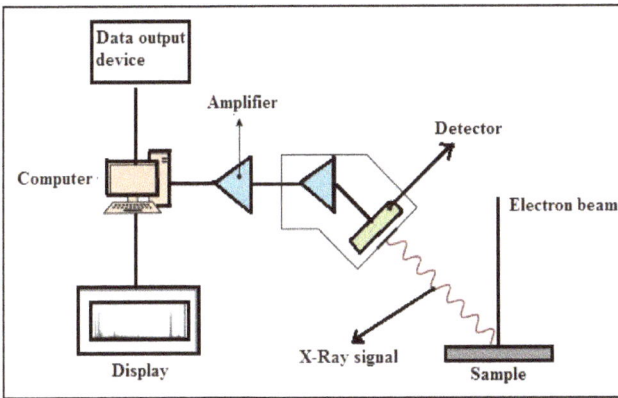

Figure 1.15 *schematic diagram of EDS. (Courtesy:*

Figure 1.15 shows the schematic diagram of the electron dispersive X-ray spectrometer. The main parts are the electron beam source; sample and detector. The electrons from the source reach the sample and produce characteristic X-rays. The signal which reaches the detector is amplified using amplifier circuits and recorded using computers. Overlapping of X-ray emission peaks Mn K_β and Fe K_α takes place and the accuracy of the measured composition is affected. The probability of the X-rays escaping the specimen, and being available for detection depends on the energy of the X-ray, the composition, and density of the material it has to pass through to reach the detector.

1.9.5 Electrical characterizations

The electrical characterizations are studied by investigating the dielectric nature and the change in the polarization with the change in the electric field (P-E hysteresis). In ferroelectric materials, the presence of electric dipoles is responsible for the polarization of the material. Similarly, the dielectric nature of the material is studied by measuring the change in the capacitance with the change in the frequency.

1.9.5.1 Dielectric characterizations

The dielectric characterization is studied by measuring the capacitance for the different values of the frequency. The pellet is placed in between two plates, which acts as a capacitor. Using an LCR circuit, the value of the capacitance is measured for different values of the frequency by measuring the voltage dropped across the pellet capacitor. As the reactance of the capacitor is given by $X_c = \frac{1}{2\pi f C}$, the voltage drop across the capacitor decreases with the increase in the frequency. The capacitance of the parallel plate capacitor is given by $C = \frac{A\epsilon}{d}$, hence the dielectric constant can be elucidated.

1.9.5.2 P-E hysteresis Loop tracer

Ferro electricity is the phenomenon, in which some materials exhibit spontaneous electric polarization, even in the absence of any externally applied field. Ferroelectric domains are the small regions in the ferroelectric crystals having polarization in the same direction. In a domain, all the electric dipoles are aligned in the same direction. There are many tiny regions in a crystal known as domains. The domain is polarized during the application of the electric field and acquires electric dipole moment. A very strong applied field would lead to the reversal of the polarization in the domain, known as domain switching. The polarization reversal is observed by measuring the ferroelectric hysteresis. The instrument used to observe P-E characterization is shown in Figure 1.16, P-E hysteresis loop tracer; NVIS 6108 at NIT, Trichy, Tamilnadu; India. The P-E hysteresis is observed from the NIT, trichy for all the samples prepared.

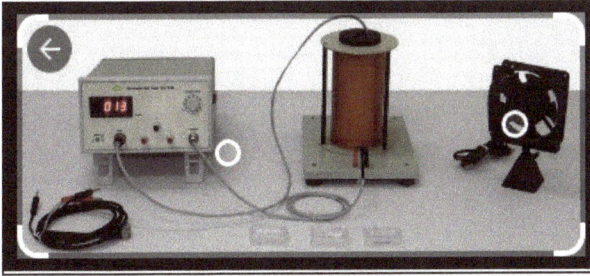

Figure 1.16 PE hysteresis loop tracer; NVIS 6108

1.9.6   M-H hysteresis (Magnetic characterization)

Magnetic hysteresis is the magnetic characterization of the ferromagnetic material in which the magnetization of the material is measured with the application of the magnetic field. When an external magnetic field is applied to a ferromagnetic material such as iron and the atomic dipoles align themselves with it. After the material has become magnetized, it retains the magnetization even when the field is removed. Once the material is magnetized, the magnet will retain magnetization. To demagnetize the material application of heat or a magnetic field in the opposite direction is done. The magnetic hysteresis curve (M-H) is similar to the electric hysteresis curve (P-E). The saturation is magnetization in M-H whereas polarization in P-E. The fields are increased, decreased, and reversed, and the magnetization or polarization is observed. The magnetization M for zero fields is known as the remnant field. The reverse field applied to demagnetize the material is coercivity. *A* small continuous, random jump in magnetization curve of the hysteresis is present in materials with crystallographic defects and dislocations known as Barkhausen jumps [*Chikazumi, Sōshin, 1997*].

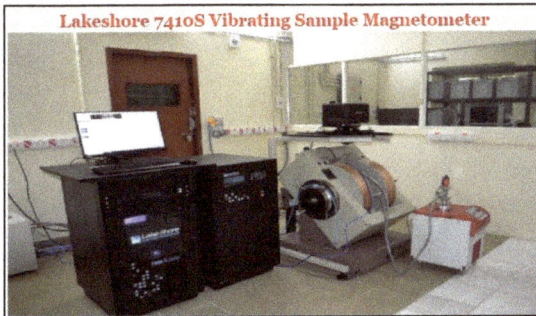

Figure 1.17 Vibrating Sample Magnetometer

The instrument used to observe the M-H characterization is shown in Figure 1.17, Lakeshore 7410S Vibrating Sample Magnetometer (VSM) at IIT Chennai; Tamilnadu, India.

1.10 Methodologies used

JANA 2006 is used for the XRD data refinement and structure analysis. Maximum Entropy Method (MEM) for electron/charge density mapping. (PRIMA and VESTA are the software used in the MEM technology.

1.10.1 Rietveld refinement: Powder XRD refinement

Rietveld refinement [Rietveld, 1967] is a technique described by Hugo Rietveld to characterize the crystalline materials from the neutron and X-ray diffraction of powder samples. The XRD of powder data show peaks in intensity at some positions called Bragg peaks. The Rietveld method is used to refine a theoretical profile from the calculated data until it matches the measured profile from the observed data. The Rietveld method fits a calculated profile (including all structural and instrumental parameters) to experimental data. The least-squares method is used in the process of refinement. The process involves the initial approximation of free parameters; peak shape, unit cell dimensions, and coordinates of all atoms in the crystal structure. In this procedure crystal structure of a powder material can be refined from PXRD data. The Rietveld method is a powerful technique that started a remarkable era for PXRD and materials science in general.

1.10.1.1 Refinement procedure

Rietveld refinement [Rietveld, 1969] method is based on the principle of minimizing a function M which is the difference between the experimentally observed profile y(obs) and theoretically calculated profile y(cal). The function M is given by,

$M = \left\{ \sum_i W_i \; y_i^{obs} - \frac{1}{c} \; y_i^{cal} \right\}^2$ where Σ is the sum of independent observations, W; statistical weight, C; an overall scale factor

JANA 2006 software [Petricek et al., 2014] is executed for the profile refinement based on this refinement. The problem is not linear in the involved parameters, approximate values of all the parameters are given in the initial refinement cycles. These parameters are refined in the subsequent cycles of refinement until a convergence is reached [Petricek et al., 2014]. The quality of the refinement between the calculated and observed profiles are estimated by the residual R-factor which is given by,

$R = \frac{\Sigma|F_{obs}-F_{cal}|}{\Sigma|Fobs|}$ F_{obs} and F_{cal} are the observed and calculated structure factors. The R-value indicates the quality of fitting between observation and constructed model. R_P and R_{obs} values indicate with the accuracy of theoretically built pattern which fits with the experimental data. The parameters that can be refined, used in the powder diffraction data are more than that of the single-crystal structure refinement. When the prepared sample has two or more phases, both the phases are refined simultaneously and the phase ratio is also refined which gives the ratio of the phases present.

1.10.1.2 Peak shape

The profile of a single diffraction peak is assumed to be Gaussian due to the combined effects of the X-ray beam, the experimental arrangement, the sample size, and its shape. The equation of XRD profile y_i at the position $2\theta_i$ is given by,

$$y_i = \left[\frac{-4\ln 2}{H_k{}^2}(2\theta_i - 2\theta_k)^2\right],$$ where I_k is calculated intensity of the Bragg reflection

H_k is full width at half maximum (FWHM) $2\theta_k$ is the calculated position of the Bragg peak. The finite slit heights with finite sample heights shows some asymmetry in the diffraction peaks at low scattering angles. A vertical divergence effect is observed where the maximum of the diffraction peak to shifts to lower angles but does not affect the integrated peak area [Klug et al., 1959]. In Rietveld [Rietveld, 1969] refinement, in addition to the Gaussian shape function, the pseudo-Voigt function of Thompson, Cox & Hastings [1987] and Lorentz function are also used. A good approximation to an asymmetric peak is obtained by introducing the correction factor in the equation given below.

$$A_s = 1 - \left[\frac{sP(2\theta_i-2\theta_k)^2}{\tan\theta_k}\right]$$

where P is an asymmetric parameter and s takes different values as, +1, 0, -1 depending on different $(2\theta_i - 2\theta_k)$ being positive, 0, negative. At a particular position, more than one peak may contribute to the profile. The intensity is the sum of all the reflections at the given position $2\theta_i$.

1.10.1.3 Peak width

Particle size plays a role in the peak broadening which is expressed with the angular dependence of the half widths of the Bragg peaks, given by the formula [Caglioti et al., 1958], $H_k{}^2 = U\tan^2\theta_k + V\tan^2\theta_k + W$ where, U, V & W are half-width parameters. By measuring the half-width H_k of selected single diffraction peaks and finding least-squares fit to these observed quantities by the equation shown above, the approximate values of these parameters are found.

1.10.1.4 Preferred orientation

The crystallite tends to align its normal along with the axis of the sample holder in polycrystalline materials. The equation for the intensity of the diffracted peak is predicted by considering the crystallites to be in random orientation. The intensity for predicted random distribution will vary and hence it should be corrected. The corrected intensity for the preferred orientation is given by,

$I_{corr} = I_{obs} \exp(-G\alpha^2)$ where, α is the acute angle between the scattering vector and the normal, G; preferred orientation parameter and I_{obs}; observed intensity.

1.10.1.5 Background function

The background, y_{bi} at step i, is approximated over a finite sum of Legendre polynomials, $F_j(x_i)$ [Abramowitz and Stegun, 1964], orthogonal relative to integration over the interval [-1, 1] is $y_{bj} = \sum_{j=0}^{n} b_j F_j(x_i)$ Fj(xi)''s for j=2 are calculated from Fj–1(xi) and Fj–2 (xi) using the relation,

$F_j(x_i) = \left[\frac{2j \pm 1}{j} x_i F_{j \pm 1}(x_i)\right]$ with the values $F_0(x_i)$=1 and $F_1(x_i)$=x_i. The coefficients, bj, are called background parameters to be refined by Rietveld [Rietveld, 1969] technique, and the variable, xi is normalized between -1 and 1.

1.10.1.6 XRD powder profile refinement using JANA 2006

Rietveld [Rietveld, 1969] refinement was carried out for all the synthesized samples using the software JANA 2006 [Petricek et al., 2014]. It is standard free software for structural refinement of the samples based on the X-ray diffraction data. The observed diffraction patterns from the data are compared with the theoretically constructed profiles using pseudo-Voigt [Wertheim, 1974], profile shape functions [Thompson, 1987], and Gaussian FWHM parameters in Rietveld [Rietveld, 1969] refinement using JANA 2006 [Petricek et al., 2014].

The profile symmetry is also introduced by using the Simpson rule of integration given by Howard [1982], which includes profile shape function with various coefficients and peak shift. March-Dollase function [March, 1932, Dollase, 1986] is used to incorporate corrections for preferred orientation in the software JANA 2006 [Petricek et al., 2014]. The theoretical profiles are constructed and compared with the observed profiles from this process of refinement. The structure factors from the Rietveld [Rietveld, 1969] refinement have been utilized for the charge density studies. The PXRD profile refinement using JANA 2006 is shown in Figure 1.18.

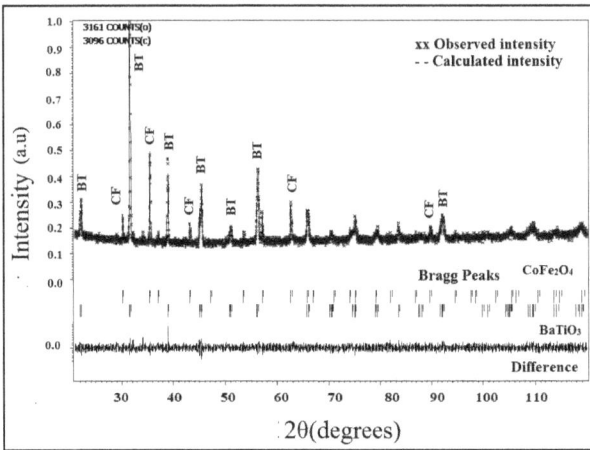

Figure 1.18 *Refined profile of a powder XRD.*

1.10.2 Maximum entropy method: Charge density analysis

The aim of the present work is a deep analysis of the precise electronic structure of the $BaTiO_3$; AFe_2O_4 composite systems where A is Ni, Co, Zn, and Mg and the bonding interaction between the constituent atoms. This aim is fulfilled successfully by adapting the maximum entropy method (MEM) [Collins, 1982]. The importance of electron density studies in structure analysis, the formalism of MEM method [Collins, 1982], the principle, MEM [Collins, 1982] methodology is followed in this work.

1.10.2.1 Electron density

The measure of the probability of an electron being present at a specific location is called electron density. The electron density is high around the atoms. An electron can be imagined as a stationary wave or a cloud of negative charges. Formulation of a method to measure the charge density in a crystalline system from its X-ray diffraction pattern was done by Debye and Scherrer [Debye et al., 1918]. The electron density is considered as a periodic function of position in a crystal, attaining a maximum value at a point at the presence of the atom and dropping to a minimum value in the region between the two atoms [Cullity and Stock, 2001]. **The electron density function $\rho(r)$ is the function which represents the density of the electrons in a volume and ρdr is the probability of finding an electron in a small volume element dr.**

The electron density distribution of a molecule is the probability distribution that clearly describes the manner in which the electronic charges are distributed throughout the real space in the field exerted by the nuclei. For the detailed investigation of bonding, electron density distribution analysis is essential [Coppens, 1989]. The most commonly adopted techniques of electron density analysis are the estimation of electron density maps and least-squares fitting of parameterized analytical functions to the observed structure factors. The 3-dimensional and 2-dimensional maps are obtained by refined structure factors. The knowledge of various phases present in the observed X-ray patterns is essential for electron density analysis [Coppens, 1979].

1.10.2.2 Structure factor

The structure factors are important to derive the charge density inside the unit cell because the charge density is the Fourier transform of structure factors. The structure factor Fhkl is the resultant of j waves scattered in the direction of reflection hkl by j atoms in the unit cell [Stout and Jensen, 1968]. The expression for the structure factor is,

$$F_{hkl} = F_{hkl}\,(_{hkl}) = \sum_j f_j \, \exp[2\pi i \,(hx_j + ky_j + lz_j)]$$

$$= \sum_j f_j \, \cos[2\pi i \,(hx_j + ky_j + lz_j)] + i \sum_j f_j \, \sin[2\pi i \,(hx_j + ky_j + lz_j)]$$

$$= A_{hkl} + iB_{hkl}$$

Sum of the structure factor is taken over all atoms in the unit cell, xj, yj, zj are the atomic coordinates of the jth atom, fj is the scattering factor of the jth atom and hkl is the phase of the diffracted beam. The scattering factor fj is the ratio of the amplitude of radiation scattered from the atom and the amplitude of radiation scattered from the single electron. The structure factor describes the way in which an incident X-ray is scattered by all the atoms in the unit cell by considering the various scattering power of the elements through the term fj. Due to the spatial distribution of the atoms in the unit cell, there will be a phase difference in the scattering amplitudes from the two atoms. This phase difference is taken into account by complex exponential form. When the summation over discrete atoms in the structure factor expression is,

$$F_{hkl} = \sum_j f_j \, e^{[2\pi i \,(hx_j + ky_j + lz_j)]}$$

The integration of the above equation of a continuous, cyclic electron density function, ρ is obtained that is in the form of Fourier transform,

$$F_{hkl} = \int_V \rho(x, y, z) e^{2\pi i (hx_j + ky_j + lz_j)} dV$$

This means that electron density is the Fourier transform of the structure factor. Similarly, the structure factor is the inverse Fourier transforms of electron density.

$$\rho(x, y, z) = \int_V' F_{hkl}e^{-2\pi i(hx_j + ky_j + lz_j)}dV$$

The above equation can be used in the form of a summation which is written as,

$$\rho(x, y, z) = \sum_h \sum_k \sum_l F_{hkl}e^{-2\pi i(hx_j + ky_j + lz_j)}dV$$

According to the above equation, the electron density can be calculated at any point (x, y, z) by constructing a Fourier series that has coefficients that are equal to the structure factors [Warren, 1990]. This is the basic equation of crystallography and it enables the calculation of electron density in the unit cell.

1.10.2.3 Formalism of maximum entropy method (MEM)

The maximum entropy method (MEM) [Collins, 1982] is used to map the electron density in the unit cell from the X-ray diffraction data: in three-dimension which gives the structure of the material, two-dimension which shows the interaction for the bonding qualitatively and one-dimension profile which is used for quantitative analysis of the electron density. MEM method [Collins, 1982] is used to extract the maximum amount of information from the X-ray data [Smaalen et al., 2009]. Initially, MEM [Collins, 1982] image technique was employed in radio astronomy. Later, it was found applications in image processing, the analysis of any type of spectroscopic or diffraction data. MEM [Collins, 1982] is now vastly used in crystallographic applications and to determine the most probable electron density distribution in the unit cell of a particular system under investigation, and locating atoms, and the determination of accurate charge density and chemical bonding. **Precise electron density maps are obtained using MEM [Collins, 1982] method. The resolution of the MEM [Collins, 1982] electron density distribution is much higher than the conventional Fourier transform [Sakata et al., 1990].** In X-ray diffraction, the experimental structure factors are applied to the inverse Fourier transform to yield the charge density. These limitations of measurement errors for the structure factors are overcome by the MEM method [Collins, 1982] of inferring the unobserved structure factors from the experimentally observed structure factors and maximizing the entropy [Saravanan et al., 2012]. The overall intensities of each reflection are evaluated from the experimentally observed X-ray diffraction patterns using the results retrieved from the Rietveld [Rietveld, 1969] refinement. By combining the Rietveld [Rietveld, 1969] refinement and MEM method [Collins, 1982], a new sophisticated technique of structure refinement in the charge density studies was evolved [Takata et al., 1995]. This Rietveld [Rietveld, 1969] and MEM [Collins, 1982] combinational analysis is basically an iterative procedure that can finally provide a better structural model [Takata et al., 2001]

1.10.2.4 Principle of MEM

The principle of MEM [Collins, 1982] is to obtain accurate electron density consistent with the experimental structure factors, and to leave the uncertainties to a minimum. The theory of the maximum entropy method (MEM) [Collins, 1982] is understandable with some equations which are similar to the equations in statistical thermodynamics because the information entropy and statistical entropy deals with the concept of most probable distribution. In statistical thermodynamics, the distribution of the particles in phase space is considered whereas, in the information theory, the distribution of numerical quantities over the ensemble of pixels is considered. The probability of the distribution of N identical particles over m boxes each containing ni particles is given by,

$$P = \frac{N!}{n_1!n_2!n_3!\dots n_m!}$$

According to statistical thermodynamics, entropy is defined as lnP. The numerator in the above equation is constant hence the entropy is written as,

$$S = - \sum_i n_i ln\, n_i$$

when there is a prior probability qi for the ith box to contain ni number of particles. This can be expressed by,

$$P = \frac{N!}{n_1!n_2!n_3!\dots n_m!}\ q_1^{n1} q_2^{n2} q_3^{n3} \dots q_m^{nm}$$

the expression for entropy,

$$S = - \sum_i n_i ln\, n_i + \sum_i n_i ln\, q_i$$
$$= - \sum_{i=1}^{m} n_i\, ln \frac{n_i}{q_i}$$

By using the equation the MEM method Collins [1982] which is expressed in information entropy of the electron density distribution as the sum over M grid points in the unit cell with the entropy formula [Jaynes,1968].

$$S = - \sum \rho'(r)\, ln\frac{\rho(r)'}{\tau(r)'}$$

The probability $\rho'(r)$ and prior probability $\tau'(r)$ is related to the electron density inside the unit cell as,

$$\rho'(r) = \frac{\rho(r)}{\sum_r \rho(r)}\ \text{and}\ \tau'(r) = \frac{\tau(r)}{\sum_r \tau(r)}$$ where $\rho(r)$ is the electron density at a certain fixed r in a unit cell and $\tau(r)$ is the prior electron density. The entropy is maximized with a constraint, which is given by,

$$C = \frac{1}{N} \sum \frac{|F_{cal}(K) - F_{obs}(K)|^2}{\sigma^2(K)}$$

Here, N is the number of reflections accounted for MEM [Collins, 1982] analysis, (k) is the standard deviation of the observation, and calculated structural factor Fcal(k) is expressed as

$$F_{cal}(k) = V \sum \rho(\mathbf{r}) \exp(-2\pi i k . \mathbf{r}) \, dV \quad \text{where V is the volume of the unit cell.}$$

The constraint C is sometimes termed as a weak constraint in which the calculated structure factors agree well with the observed ones when C becomes unity. The structure factor given in the equation is the Fourier transform of the electron density inside the unit cell. The structure factors are normally written without introducing the atomic form factors. It should be stressed here that it would be an assumption to use the atomic form factors in the formulation of structure factors. The equation (1.38) confirms that it is possible to allow any type of deformation in the electron densities in real space as long as information concerning such a deformation is included in the observed data. In order to constrain the function C to be unity during entropy maximization, Lagrange"s method of the undetermined multiplier is used.

Then, $Q = S - \dfrac{\lambda}{2} C = -- \sum \rho'(\mathbf{r}) \ln\dfrac{\rho(r)'}{\tau(r)'} -- \dfrac{\lambda}{2}\dfrac{1}{N} \sum \dfrac{|F_{cal}(K) - F_{obs}(K)|^2}{\sigma^2(K)}$ when dρ/dQ = 0, and using the approximation, $\ln x = x-1$ then the electron density,

$$\rho(r)_i = \tau(r)_i \exp\left\{ \left(\dfrac{\lambda F_{000}}{N}\right)\left[\sum \dfrac{1}{\sigma(k)^2}\right] |F_{cal}(K) - F_{obs}(K)| \exp(-2\pi j k . r)\right\}$$

where, $F_{000}=Z$, the total number of electrons in a unit cell. Equation (1.40) cannot be solved as it is, since Fobs(k) is defined on $\rho(r)$.

In the Fourier summation, the amplitudes of the unobserved reflections are assumed to be zero, but MEM [Collins, 1982] technique provides the most probable values. MEM [Collins, 1982] has many advantages compared to the conventional Fourier method in electron density calculations are,

(1) Provides an explicit formulation for the actual electron density.

(2) Leads to the least biased calculation.

(3) Performs accurately even with a limited number of information.

(4) Unobserved reflections can be simulated.

(5) Precise electron density maps can be obtained.

(6) The existence of bonding electrons can be clearly visualized.

These are the advantages over other methods, considerable research on the charge density analysis using the MEM [Collins, 1982] method is going on successfully.

1.10.2.5 MEM methodology

A three-dimensional description of the electron density in a selected crystal structure can be evaluated from the X-ray diffraction patterns because X-rays are scattered from the electron clouds of atoms in a crystal lattice. The structure factors retrieved from the Rietveld [Rietveld, 1969] refinement are utilized for the evaluation and visualization of the charge density of the prepared systems. In this work, the electronic structure and the spatial electron density inside the unit cell have been successfully executed using sophisticated computer programs. All the data sets are refined using the software PRIMA (PRactice Iterative MEM analysis) [Izumi et al., 2002]. PRIMA is software for MEM [Collins, 1982] analysis to calculate the electron densities from the X-ray diffraction data.

The unit cell was partitioned into $64 \times 64 \times 64$ pixels for cubic systems and the prior electron density in each of the pixels was uniformly fixed as $\mathbf{Z/a_0^3}$, where Z; total number of electrons in the unit cell, a_0^3; volume of the unit cell. The Lagrangian multiplier in each case is selected suitably such that, the convergence criterion C becomes unity after a minimum number of iterations. The clear visualization of electron density in 3D and 2D levels was plotted using VESTA (**V**isualization for **E**lectron and **ST**ructural **A**nalysis) [Momma, 2008] software package. VESTA [Momma, 2008] represents the crystal structure by using various models like ball and stick, space-filling, polyhedral, and wireframe. With all the advanced features, the software VESTA [Momma, 2008] has been effectively utilized in this work for electronic structure studies in three-dimensional, two-dimensional, and one-dimensional.

1.10.3 Tauc Plot: Bandgap from UV-Visible spectrum

UV-Vis absorption analysis is the most frequently used method for optical characterization of the material under investigation. The optical band gap is evaluated from the data of the absorption spectrum using the method proposed by Wood and Tauc [Wood and Tauc, 1972]. The optical bandgap depends on the absorbance (α) and photon energy (hν) by the following relation, $\alpha h\nu = (h\nu - Eg)^n$ where $\boldsymbol{\alpha}$ is the absorbance, hν is photon energy, **Eg** is the optical bandgap, **A** is a constant and **n** is an index which takes different values for different transition types of transitions, viz., **n=1/2, for direct allowed transition, n=3/2 for direct forbidden transition, n=2 for indirect allowed transition and n=3 for indirect forbidden transition.** For direct bandgap materials, the equation is $\alpha h\nu = (h\nu - Eg)^{1/2}$, squaring the equation; $\boldsymbol{(\alpha h\nu)^2 = A(h\nu - Eg)}$, $\boldsymbol{(\alpha E^2) = Ah\nu - AEg}$. The equation resembles the equation of a straight line, $y = mx + C$, Comparing the equations $y = (\alpha E^2)$ If $y=0$, then the equation becomes, $mx + C = 0$. $A\, h\nu - Eg = 0$. In the equation, the constant A cannot be equal to zero, then, $h\nu - Eg = 0$ and $Eg = h\nu$. Using UV-vis absorption data, a graph can be drawn by taking energy (hν) in the x-axis, and $(\alpha h\nu)2$ in the y-axis. From this Tauc plot, the extrapolation of the tangent of the linear portion of the curve to the x-axis will give the value of optical band gap Eg.

1.10.4 Grain size: GRAIN software

The average grain size is evaluated using full width at half maximum (FWHM) of the powder XRD peaks using the Scherrer formula [Cullity and Stock, 2001] which is given by, $t = 0.9\lambda/\beta cos\theta$ where, λ is the wavelength of X-ray, which is 1.54056 Å, β is the Full width at half maximum (FWHM) in radians, and θ; Bragg angle, t; average grain size, which is the average size of coherently diffracting domains. GRAIN software [Saravanan, personal communication] is used to examine the average grain size of all the prepared samples.

References

• Abramowitz M and Stegun I. A, Handbook of Mathematical Functions, National Bureau of Standards, (1964).

• Aubert, A. "Enhancement of the Magnetoelectric Effect in Multiferroic CoFe2O4/PZT Bilayer by Induced Uniaxial Magnetic Anisotropy". IEEE Transactions on Magnetics. 53 (11): 1(2017) https://doi.org/10.1109/TMAG.2017.2696162

• Aubert, A. "Uniaxial anisotropy and enhanced magnetostriction of CoFe2O4 induced by reaction under uniaxial pressure with SPS". Journal of European Ceramic Society 37 (9): 3101(2017) https://doi.org/10.1016/j.jeurceramsoc.2017.03.036

• Babu Naidu K. C., Sofi Sarmash T., Maddaiah M., Gurusampath Kumar A., Jhansi Rani D., Sharon Samyuktha V., Obulapathi L., Subbarao T., Journal of Ovonic Research 11(2), 79 (2015)

• Ben Ali M., El Maalam K., El Moussaoui H., Mounkachi O., Hamedoun M., Masrour R., Hlil E.K., Benyoussef A., Effect of zinc concentration on the structural and magnetic properties of mixed Co-Zn ferrites nanoparticles synthesized by sol/gel method. J.Magn. Magn. Mater 398, 20 (2016) https://doi.org/10.1016/j.jmmm.2015.08.097

• Brabers V. A. M. (Elsevier), Handbook of Magnetic Materials, Vol. 8 189, (1995) https://doi.org/10.1016/S1567-2719(05)80032-0

• Carter, C. Barry; Norton, M. Grant Ceramic Materials: Science and Engineering 212 (2007). ISBN 978-0-387-46270-7.

• Castel V., Brosseau C., and Ben Youssef J. Magnetoelectric effect in BaTiO3/Ni particulate nanocomposites at microwave frequencies J. Appl. Phys. 106 064312; doi:

10.1063/1.3225567 (2009) https://doi.org/10.1063/1.3225567

- Chandra Babu Naidu K., Sofi Sarmash T., Narasimha Reddy V., Maddaiah M., Sreenivasula Reddy P., Subbarao T., Journal of The Australian Ceramic Society 51(1), 94 (2015).

- Cheong S., Mostovoy M., Multiferroics: a magnetic twist for ferroelectricity, Nat. Mater. 6 13 (2007). https://doi.org/10.1038/nmat1804

- Chikazumi, Sōshin. Physics of ferromagnetism (2nd ed.) Oxford: Oxford University Press. (1997)

- Chung Ming Leung, Jiefang Li, Viehland D, Zhuang X A review on applications of magnetoelectric composites: from heterostructural uncooled magnetic sensors, energy harvesters to highly efficient power converters J. Phys. D: Appl. Phys. 51 263002 (2018) https://doi.org/10.1088/1361-6463/aac60b

- Collins D. M, Nature. 298, 49 (1982). https://doi.org/10.1038/298049a0

- Coppens P, Guru Row T. N, Leung P, Stevens E. D, Becker P. J, Yang Y, Acta Cryst. A35, 63 (1979). https://doi.org/10.1107/S0567739479000127

- Coppens P, J. Phys. Chem. 93, 7979 (1989). https://doi.org/10.1021/j100361a006

- Cullity B. D, Stock S. R, Elements of X-ray diffraction, Pearson education. 3rd ed. Prentice Hall, Upper Saddle River, 558 (2001).

- Debye P, Scherrer P, Phys. Zeit. 19, 474 (1918).

- Dollase W. A, J Appl Crystallogr. 19, 267 (1986) https://doi.org/10.1107/S0021889886089458

- Dong S., Zhai J., Li J. F., Viehland D., Priya S., Energy harvesting from ambient low-frequency magnetic field using magneto-mechano-electric composite cantilever Appl. Phys. Lett. 93 103511 (2008) https://doi.org/10.1063/1.2982099

- Elhajjar, Rani; La Saponara, Valeria; Muliana, Anastasia, eds. Smart Composites: Mechanics and Design (Composite Materials). CRC Press. ISBN 978-1-138-07551-1. (2017)

- Everhart, T. E.; Thornley, R. F. M. "Wide-band detector for micro-microampere low-energy electron currents" (PDF). Journal of Scientific Instruments. 37 (7): 246 (1960). Bibcode:1960JScI...37..246E. https://doi.org/10.1088/0950-7671/37/7/307

- Fazeli, Mahyar; Florez, Jennifer Paola; Simão, Renata Antoun. "Improvement in adhesion of cellulose fibers to the thermoplastic starch matrix by plasma treatment modification". Composites PartB: Engineering. 163 207 (2019) https://doi.org/10.1016/j.compositesb.2018.11.048

• Fazeli, Mahyar; Keley, Meysam; Biazar, Esmaeil "Preparation and characterization of starch-based composite films reinforced by cellulose nanofibers". International Journal of Biological Macromolecules. 116 272 (2018) https://doi.org/10.1016/j.ijbiomac.2018.04.186

• Goldstein, G. I.; Newbury, D. E.; Echlin, P.; Joy, D. C.; Fiori, C.; Lifshin, E. Scanning electron microscopy and x-ray microanalysis. New York: Plenum Press (1981). ISBN 978-0-306-40768-0. https://doi.org/10.1007/978-1-4613-3273-2

• Gubin, Sergei P; Koksharov, Yurii A; Khomutov, G B; Yurkov, Gleb Yu "Magnetic nanoparticles: preparation, structure and properties". Russian Chemical Reviews. 74 (6): 489 (30 June 2005). doi:10.1070 /RC2005 v074n06 ABEH000897 https://doi.org/10.1070/RC2005v074n06ABEH000897

• Gul I.H., Pervaiz E., Comparative study of $NiFe_2-xAlxO_4$ ferrite nanoparticles synthesized by chemical co-precipitation and sol-gel combustion techniques. Mater. Res. Bull. 47, 1353-1361(2012) https://doi.org/10.1016/j.materresbull.2012.03.005

• Hosni "Semi-hard magnetic properties of nanoparticles of cobalt ferrite synthesized by the co-precipitation process". Journal of Alloys and Compounds 694: 1295(2016) https://doi.org/10.1016/j.jallcom.2016.09.252

• Hou Y. H., Zhao Y. J., Liu Z. W., Yu H. Y., Zhong X. C., Qiu W. Q., Zeng D. C. and Wen L.S., "Structural, Electronic and Magnetic Properties of Partially Inverse Spinel $CoFe_2O_4$: a First-Principles Study", J. Appl. Phys 43 445003 (2010) https://doi.org/10.1088/0022-3727/43/44/445003

• Howard C. J, J Appl Crystallogr. 15, 615 (1982). https://doi.org/10.1107/S0021889882012783

• Izumi F, Dilanien R.A, Recent Research Developments in Physics Part II 3 Transworld Research Network. Trivandrum 699 (2002).

• Jaynes E. T, IEEE Trans Syst Sci Cybern SSC. 4, 227 (1968). https://doi.org/10.1109/TSSC.1968.300117

• Jenkins, R. A.; De Vries, J. L. Practical X-Ray Spectrometry Springer. (1982) ISBN 978-1-468-46282-1.

• Klug H. P, Alexander L. E, X-ray diffraction procedures, second edition, John Wiley New York 251 (1959).

• Marauska S., Jahns R., Kirschof C., Klaus M., Quandt E., Knöchel R., Wagner B.,

Sensor. Magnetoelectric coupling on multiferroic cobalt ferrite-barium titanate ceramic composites with different connectivity schemes Actuat. A-Phys. 189 321(2013) https://doi.org/10.1016/j.sna.2012.10.015

- Masoudpanah S.M., Seyyed Ebrahimi S.A., Derakhshani M., Mirkazemi S.M., Structure and magnetic properties of La substituted ZnFe2O4 nanoparticles synthesized by sol-gel autocombustion method. J. Magn. Magn. Mater. 370 122 (2014) https://doi.org/10.1016/j.jmmm.2014.06.062

- McEvoy, M. A.; Correll, N. "Materials that couple sensing, actuation, computation, and communication". Science. 347 (6228) 1261689 (2015) https://doi.org/10.1126/science.1261689

- Mitoseriu L., Buscaglia V., Viviani M., Buscaglia M.T., Pallecchi I., Harnagea C., Testino A., TrefilettiV.,NanniP.,SiriA.S., BaTiO3-Ni0.5Zn0.5Fe2O4 ceramic composites with ferroelectric and magnetic properties, J.Eur.Ceram.Soc.27 4379 (2007) https://doi.org/10.1016/j.jeurceramsoc.2007.02.167

- Momma K, Izumi F, VESTA: a three-dimensional visualization system for electronic and structural analysis. J. Appl. Crystallogr. 41, 653 (2008). https://doi.org/10.1107/S0021889808012016

- Narendrababu S., Bhimashankaran T., Suryanarayana S. V., Magnetoelectric effect in metal - PZT laminates. Bull. Mater. Sci. 28 419 (2005) https://doi.org/10.1007/BF02711230

- Neaton J., Ederer C., Waghmare U., Spaldin N., Rabe K., First-principles study of spontaneous polarization in multiferroic BiFeO3, Phys. Rev. B 71 014113 (2005). https://doi.org/10.1103/PhysRevB.71.014113

- Olabi "Design and application of magnetostrictive materials" (PDF). Materials & Design. 29 (2) 469 (2008) https://doi.org/10.1016/j.matdes.2006.12.016

- Petricek V, Dusek M, Palatinus L, Kristallogr Z, Crystallographic Computing System JANA2006: General features, 229, 345 (2014). https://doi.org/10.1515/zkri-2014-1737

- Rietveld H. M, Acta Crystallogr. 22, 151 (1967). https://doi.org/10.1107/S0365110X67000234

- Rietveld H. M, J. Appl. Crystallogr. 2, 65 (1969). https://doi.org/10.1107/S0021889869006558

- Saravanan R, GRAIN software, Personal communication.

- Saravanan R, Rani M. P, Metal and Alloy Bonding- an Experimental Analysis, Chapter

2 Springer-Verlag London Press, (2012). https://doi.org/10.1007/978-1-4471-2204-3

- Sato Turtelli; et al. "Co-ferrite - A material with interesting magnetic properties". Iop Conference Series: Materials Science and Engineering. 60 012020 (2014) https://doi.org/10.1088/1757-899X/60/1/012020

- Schmid H., Multi-ferroic magnetoelectrics, Ferroelectrics 162 317 (1994) https://doi.org/10.1080/00150199408245120

- Seyyed Ebrahimi S.A., Masoudpanah S.M., Effects of pH and citric acid content on the structure and magnetic properties of Mn-Zn ferrite nanoparticles synthesized by a sol-gel autocombustion method. J. Magn. Magn. Mater. 357 77 (2014) https://doi.org/10.1016/j.jmmm.2014.01.017

- Shriver, D.F.; et al. Inorganic Chemistry. New York: W.H. Freeman. (2006) ISBN 978-0-7167-4878-6

- Skoog, Douglas A.; Holler, F. James; Crouch, Stanley R. Principles of Instrumental Analysis (6th ed.). Belmont, CA: Thomson Brooks/Cole. 169 (2007). ISBN 978-0-495-01201-6

- Slonczewski J. C. "Origin of Magnetic Anisotropy in Cobalt-Substituted Magnetite" Physical Review. 110 (6): 1341(1958) doi:10.1103/PhysRev.110.1341 https://doi.org/10.1103/PhysRev.110.1341

- Smaalen S.V, Jeanette Netzel J, Phys. Scr. 79, 048304 (2009). https://doi.org/10.1088/0031-8949/79/04/048304

- Spaldin, Nicola A. Magnetic Materials: Fundamentals and Applications, 2nd Ed. Cambridge University Press 120. (2010) ISBN 9781139491556

- Stokes, Debbie J Principles and Practice of Variable Pressure Environmental Scanning Electron Microscopy (VP-ESEM). Chichester: John Wiley & Sons. (2008) ISBN 978-0470758748 https://doi.org/10.1002/9780470758731

- Stout G. H, Jensen L. H, X-ray structure determination-a practical guide. The Macmillan Company Collier-Macmillan, London, 217 (1968).

- Takada T., Bando Y., Kiyama M., Shinjo T., Hoshino, Iida S., Suginoo M. (Eds.), Proceedings of the International Conference on Ferrites, Japan, 29- 31, July 1970, University of Tokyo Press, Japan, (1971)

- Thompson P, Cox D. E, Hastings J. B, J Appl Crystallogr. 20, 79 (1987). https://doi.org/10.1107/S0021889887087090

- Ullah, Zaka; Atiq, Shahid; Naseem, Shahzad "Influence of Pb doping on structural, electrical and magnetic properties of Sr-hexaferrites". Journal of Alloys and Compounds 555: 263 (2013) doi:10.1016/j.jallcom.2012.12.061 https://doi.org/10.1016/j.jallcom.2012.12.061

• Van den Boomgaard J., Van Run A. M. J. G., Van Suchtelen J., Piezoelectric-piezomagnetic composites with magnetic effect. Ferroelectrics 14 727 (1976) https://doi.org/10.1080/00150197608236711

• Wang "Magnetostriction properties of oriented polycrystalline CoFe2O4". Journal of Magnetism and Magnetic Materials. 401: 662 (2015). doi:10.1016/j.jmmm.2015.10.073 https://doi.org/10.1016/j.jmmm.2015.10.073

• Warren B. E, X-ray diffraction, Chapter 3. Dover publications, New York (1990).

• Wertheim G. K, Butler M. A, West K. W, Buchanan D. N. E, Rev Sci Instrum. 45, 1369 (1974). https://doi.org/10.1063/1.1686503

• Wood D. L, Tauc J , Phys Rev B. 5, 3144 (1972). https://doi.org/10.1103/PhysRevB.5.3144

• Yang S., Han D., Wang Z., Liu Y., Chen G., Luan H., Bayanheshig L., Yang, Synthesis and magnetic properties of ZnFe1.97RE0.03O4 (RE¼Eu and Nd) nanoparticles. Mater. Sci. Semicond. Process 27 854 (2014) https://doi.org/10.1016/j.mssp.2014.08.032

Chapter 2

Results

Abstract

Chapter II depicts the results of the four series of magnetoelectric composite, prepared by the solid-state reaction process. $(1-x)$ $BaTiO_3$ + x MFe_2O_4 (M = Co, Zn, Ni, Mg and x = 0.2, 0.4, 0.6, 0.8). The raw profiles for various compositions in a series of the XRD data are presented. The refined profiles of the XRD data (Rietveld fitted profiles), SEM micrographs, EDS spectra, three-dimensional (3D), two-dimensional (2D), and one-dimensional (1D) electron density distribution from the (Maximum Entropy Method) MEM method are presented in this chapter. UV-Vis absorption graphs Tauc plots to find the optical bandwidth. The electrical characterization: Capacitance vs. frequency graph, P-E hysteresis graph, and the magnetic characterization; M-H hysteresis plot. Synthesis of $(1-x)BaTiO_3$ + $xNiFe_2O_4$; $(1-x)BaTiO_3$ + $xZnFe_2O_4$; $(1-x)BaTiO_3$ + $xCoFe_2O_4$ and $(1-x)BaTiO_3$ + $xMgFe_2O_4$ series were carried out by solid state sintering method. Various characterizations: structural, morphological, optical, electrical and magnetic were studied. The microscopy and the microstructure of the prepared samples is analyzed using the Scanning electron microscope (SEM) image or micrographs. The SEM micrographs for all the samples are presented. The elemental composition of all the composites is analyzed from the EDXS data. The optical characterization of the prepared samples has been carried out from the UV-Visible spectroscopy measurements. The bandgap of the samples has been evaluated from the UV-Visible data. Tauc plot is used to plot the UV-Visible data; in the procedure given by [Wood and Tauc, 1972] and from this plot, the bandgap of the sample is interpreted. Optical bandgap values of all the composites with different compositions is listed. The dielectric characterization is performed to investigate the electrical nature of the prepared samples and the capacitance versus frequency graphs for all the samples synthesized, dielectric loss characterization, the P-E hysteresis for all the samples is studied to analyze the ferroelectric nature of the samples. The magnetic characterization is analyzed for all the prepared from the VSM data. The M-H hysteresis curves for all the compositions of the composite.

Keywords

XRD, SEM, EDS, electron density distribution, Maximum Entropy Method, MEM, UV-Vis, Capacitance vs. frequency graph, P-E hysteresis graph, M-H hysteresis

2.1 Introduction

The work done includes synthesis and characterization of four series of magneto-electric ceramic composites. Synthesis of $(1\text{-}x)BaTiO_3$ + $xNiFe_2O_4$; $(1\text{-}x)BaTiO_3$ + $xZnFe_2O_4$; $(1\text{-}x)BaTiO_3$ + $xCoFe_2O_4$ and $(1\text{-}x)BaTiO_3$ + $xMgFe_2O_4$ series were carried out by solid state sintering method. Various characterizations: structural, morphological, optical, electrical and magnetic were studied. Electron density studies were analyzed in three-dimensional, two-dimensional and one-dimensional to correlate with the characterizations observed.

Powder X-ray diffraction data is used to analyze the structure of the composite. XRD data was refined by Rietveld refinement [Rietveld, 1969] using the software JANA 2006 [Petricek et. al., 2014]. The experimental XRD profiles, refined fitted profiles by Rietveld refinement [Rietveld, 1969] and the structural parameters from the refinement is given in section 2.2.

The microscopy and the microstructure of the prepared samples is analyzed using the Scanning electron microscope (SEM) image or micrographs. The SEM micrographs for all the samples are presented in section 2.3. The elemental composition of all the composites is analyzed from the EDXS data. The EDXS data for all the compositions of the composites are listed in section 2.4

The optical characterization of the prepared samples has been carried out from the UV-Visible spectroscopy measurements. The bandgap of the samples has been evaluated from the UV-Visible data. Tauc plot is used to plot the UV-Visible data; in the procedure given by [Wood and Tauc, 1972] and from this plot, the bandgap of the sample is interpreted. Optical bandgap values of all the composites with different compositions are given in section 2.5.

The dielectric characterization is performed to investigate the electrical nature of the prepared samples and the capacitance versus frequency graphs for all the samples synthesized, dielectric loss characterization, the P-E hysteresis for all the samples is studied to analyze the ferroelectric nature of the samples are given in section 2.6.

The magnetic characterization is analyzed for all the prepared from the VSM data. The M-H hysteresis curves for all the compositions of the composite are presented in section 2.7.

The electronic structural analysis, the electron density distribution, and chemical bonding are investigated using standard software MEM (Maximum Entropy Method) [Collins, 1982]. The MEM [Collins, 1982] method has been carried out using the softwares PRIMA [Izumi, 2002] and VESTA Momma, 2008]. The three-dimensional, two-dimensional and one-dimensional electron density distributions have been plotted and analyzed using the visualization software VESTA [Momma, 2008] and are presented in section 2.8.

2.2 Structural characterization: Powder X-ray diffraction

2.2.1 $(1-x)BaTiO_3 + xNiFe_2O_4$

The barium titanate-nickel ferrite composite have been synthesized by solid state reaction method for the compositions 0.2, 0.4, 0.6, and 0.8. The raw XRD profiles are presented in Figure 2.1. The refined profiles are depicted in Figure 2.2 (a) for x = 0.2, (b) for x = 0.4, (c) for x = 0.6 , and (d) for x = 0.8. The lattice parameters and volume from the Rietveld refinement are given in table 2.1.

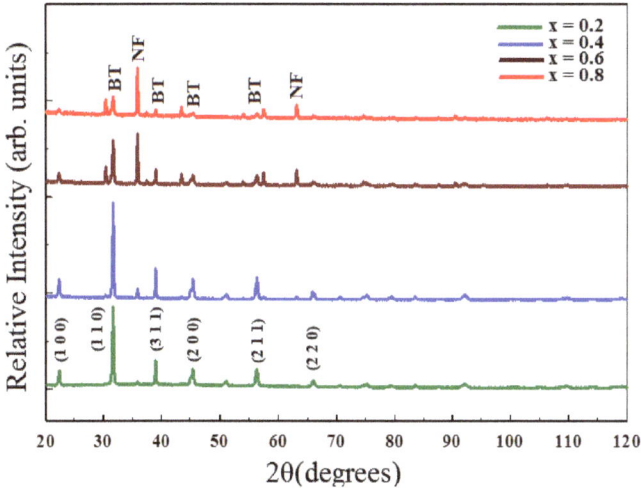

Figure 2.1 *Raw XRD profile of $(1-x)BaTiO_3 + xNiFe_2O_4$ for x = 0.2, 0.4, 0.6, and 0.8.*

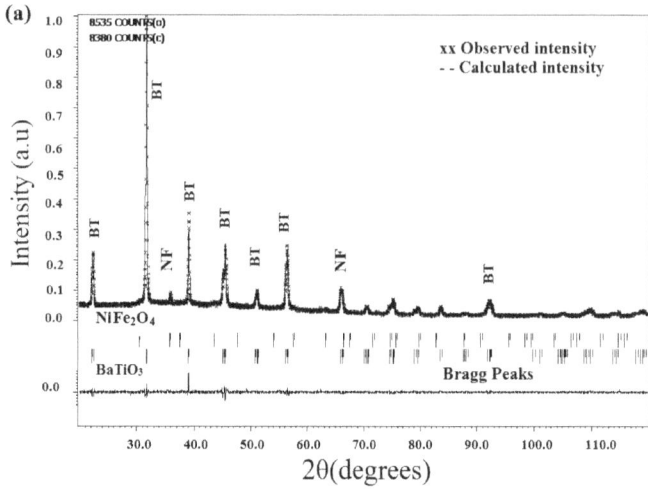

(a)

8535 COUNTS(o)
8380 COUNTS(c)

xx Observed intensity
- - Calculated intensity

Figure 2.2(a) *Fitted XRD profile of (1-x)BaTiO₃ + xNiFe₂O₄ prepared composite for composition x = 0.2.*

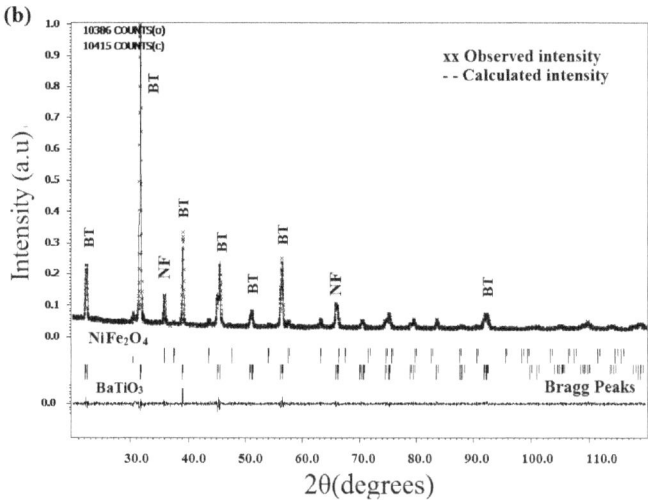

(b)

10386 COUNTS(o)
10415 COUNTS(c)

xx Observed intensity
- - Calculated intensity

Figure 2.2(b) *Fitted XRD profile of (1-x)BaTiO₃ + xNiFe₂O₄ prepared composite for composition x = 0.4.*

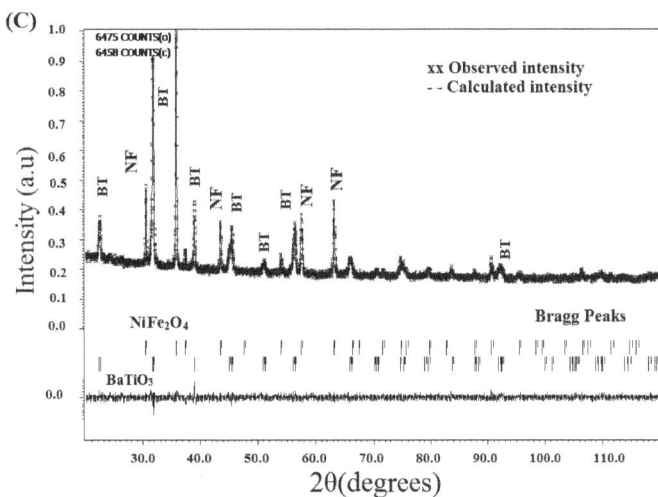

Figure 2.2(c) *Fitted XRD profile of (1-x)BaTiO₃ + xNiFe₂O₄ prepared composite for composition x = 0.6.*

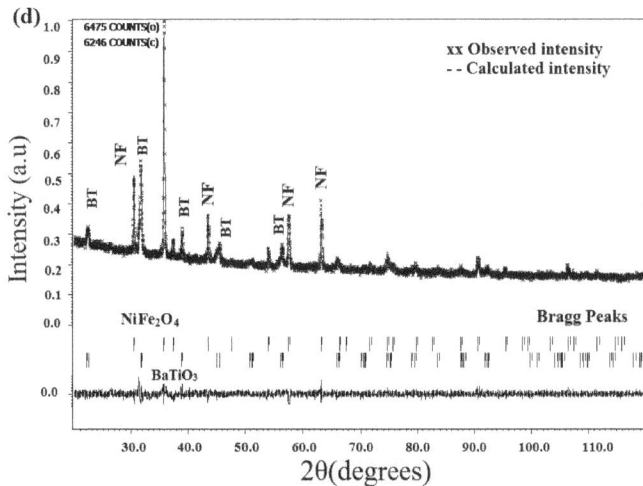

Figure 2.2(d) *Fitted XRD profile of (1-x)BaTiO₃ + xNiFe₂O₄ prepared composite for the composition x = 0.8.*

Table 2.1 *Lattice Parameters and Volume of (1-x) BaTiO₃ + x NiFe₂O₄ composite.*

Composition	Ferroelectric (BaTiO₃)		Ferrite (NiFe₂O₄)	Volume (Å³)		c/a
x	a (Å)	c (Å)	a (Å)	Ferroelectric	Ferrite	ratio
0.2	3.9985(0)	4.0318(0)	8.3233(0)	64.4609	576.6202	1.0083
0.4	4.0042(0)	4.0346(11)	8.3383(5)	64.6260	579.7415	1.0076
0.6	3.9949(0)	4.0285(0)	8.3449(8)	64.2932	581.1196	1.0084
0.8	3.9964(0)	4.0331(0)	8.3430(0)	64.4121	580.7238	1.0091

2.2.2   (1-x)BaTiO₃ + xZnFe₂O₄

Solid state reaction method is used to synthesize barium titanate-Zinc ferrite composite for the compositions 0.2, 0.4, 0.6, and 0.8. The raw XRD profiles are presented in Figure 2.3. The refined profiles are depicted in Figure 2.4 (a) for x = 0.2, (b) for x = 0.4, (c) for x = 0.6, and (d) for x = 0.8. The lattice parameters and volume from the Rietveld refinement are given in table 2.2.

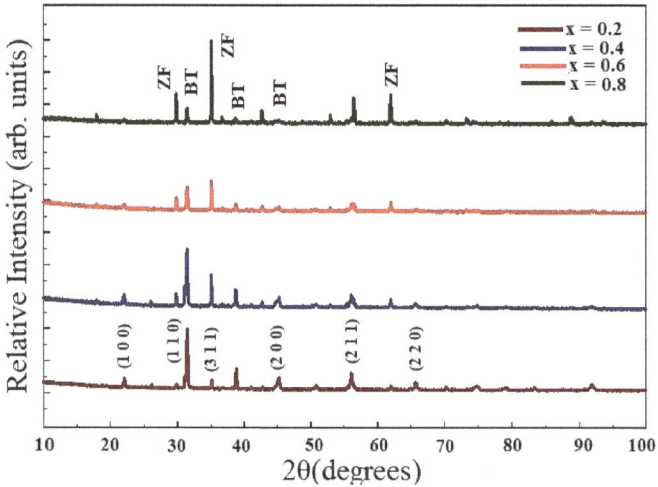

Figure 2.3 *Raw XRD profile of (1-x)BaTiO₃ + xZnFe₂O₄ for x = 0.2, 0.4, 0.6, and 0.8.*

(a)

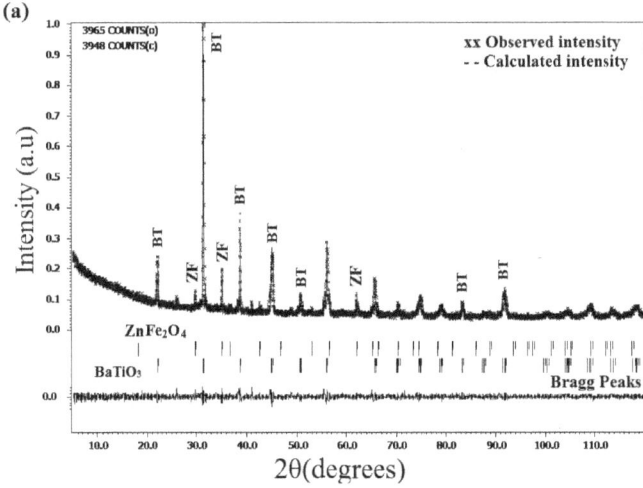

Figure 2.4(a) *Refined XRD profile of (1-x)BaTiO₃ + xZnFe₂O₄ prepared composite for composition x = 0.2.*

(b)

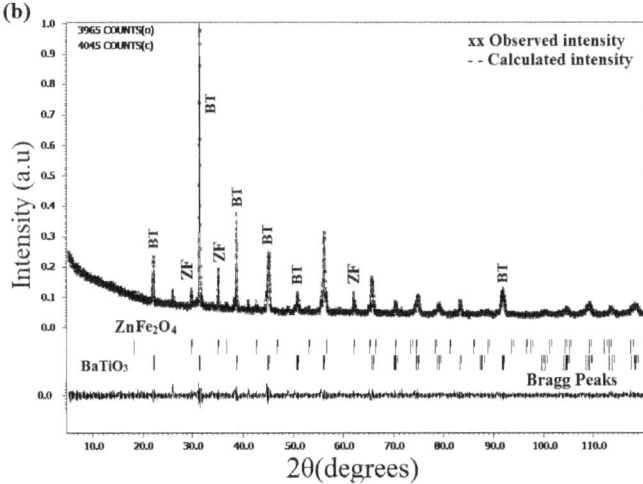

Figure 2.4(b) *Refined XRD profile of (1-x)BaTiO₃ + xZnFe₂O₄ prepared composite for composition x = 0.4.*

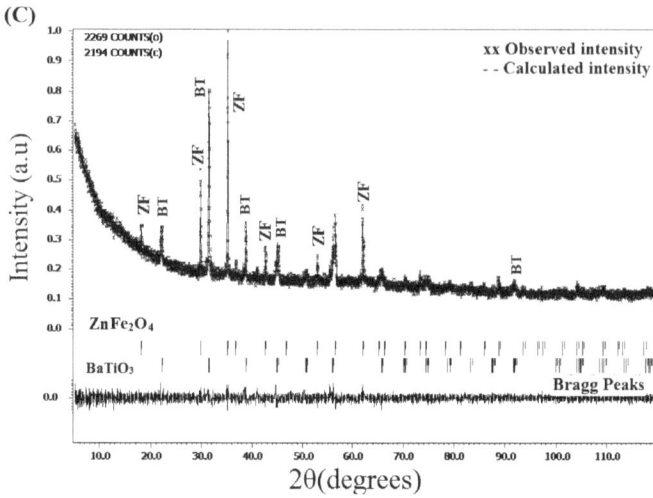

Figure 2.4(c) *Refined XRD profile of (1-x)BaTiO₃ + xZnFe₂O₄ prepared composite for composition x = 0.6.*

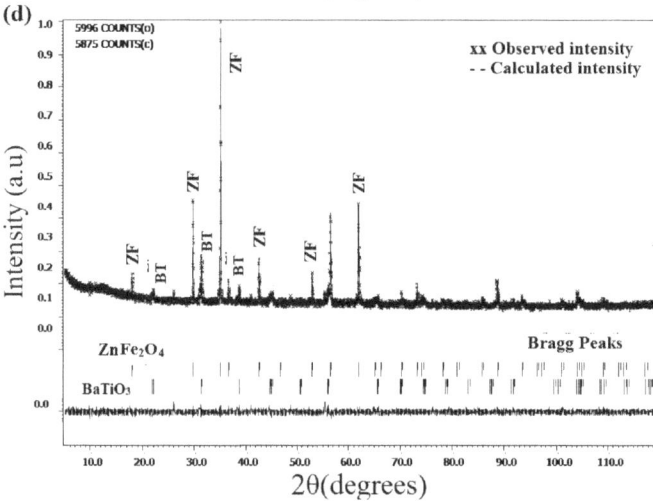

Figure 2.4(d) *Refined XRD profile of (1-x)BaTiO₃ + xZnFe₂O₄ prepared composite for composition x = 0.8.*

Table 2.2 *Lattice parameters and volume of (1-x) BaTiO₃ + x ZnFe₂O₄ composite.*

Composition	Ferroelectric (BaTiO₃)		Ferrite	Volume (Å³)		
x	a (Å)	c (Å)	(ZnFe₂O₄) a (Å)	Ferroelectric	Ferrite	c/a
0.2	4.0058(2)	4.0281(2)	8.4492(0)	64.636(5)	603.177(23)	1.0055
0.4	4.0019(0)	4.0268(0)	8.4442(0)	64.488(0)	602.100(0)	1.0062
0.6	4.0004(0)	4.0240(0)	8.4482(0)	64.398(0)	602.976(0)	1.0059
0.8	4.0070(0)	4.0314(0)	8.4424(0)	64.728(0)	601.728(0)	1.0060

2.2.3   (1-x)BaTiO₃ + xCoFe₂O₄

Solid state reaction method is used to synthesize barium titanate-cobalt ferrite composite for the compositions 0.2, 0.4, 0.6, and 0.8. The raw XRD profiles are presented in Figure 2.5. The refined profiles are depicted in Figure 2.6 (a) for x = 0.2, (b) for x = 0.4, (c) for x = 0.6, and (d) for x = 0.8. The lattice parameters and volume from the Rietveld refinement is given in table 2.3.

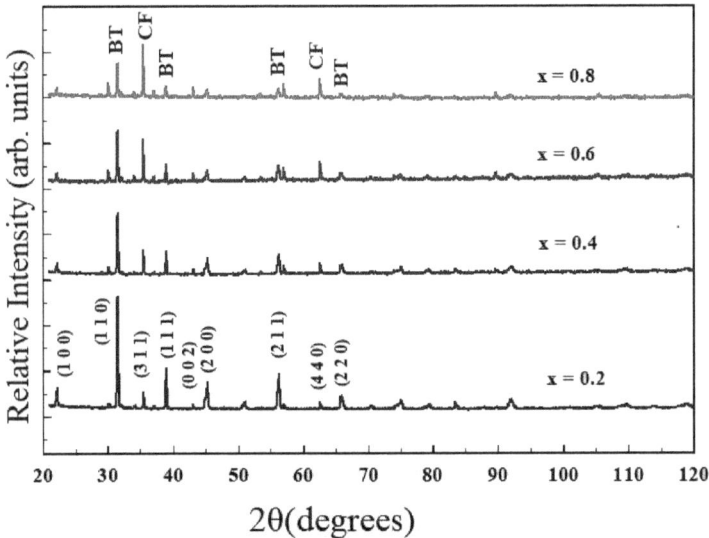

Figure 2.5 *Raw XRD profile of (1-x)BaTiO₃ + xCoFe₂O₄ for x = 0.2, 0.4, 0.6, and 0.8.*

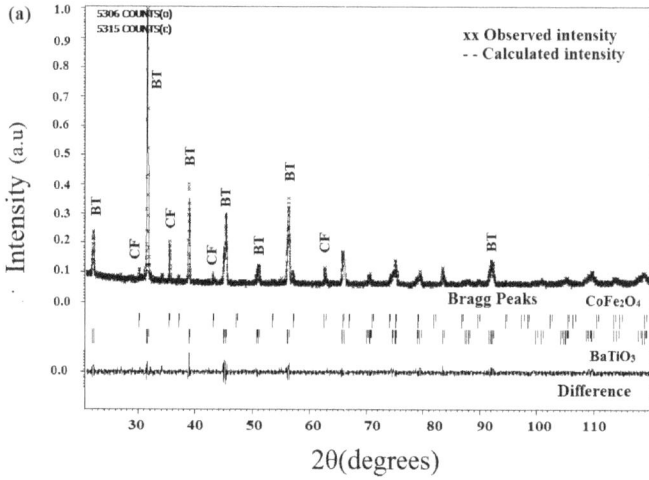

Figure 2.6(a) *Fitted XRD profile of (1-x)BaTiO₃ + xCoFe₂O₄ prepared composite for composition x = 0.2.*

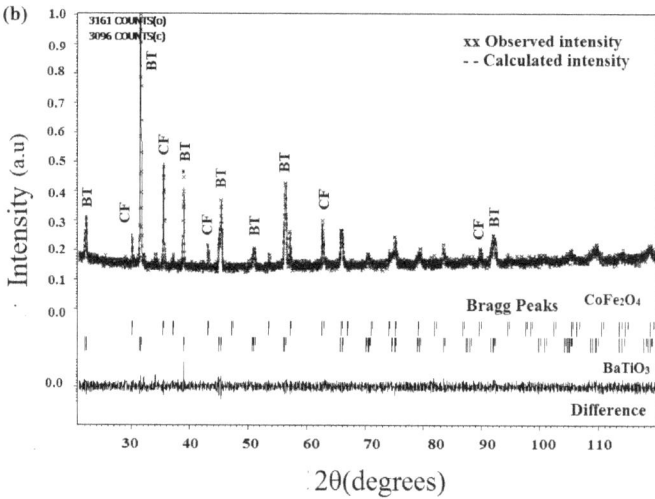

Figure 2.6(b) *Fitted XRD profile of (1-x)BaTiO₃ + xCoFe₂O₄ prepared composite for composition x = 0.4.*

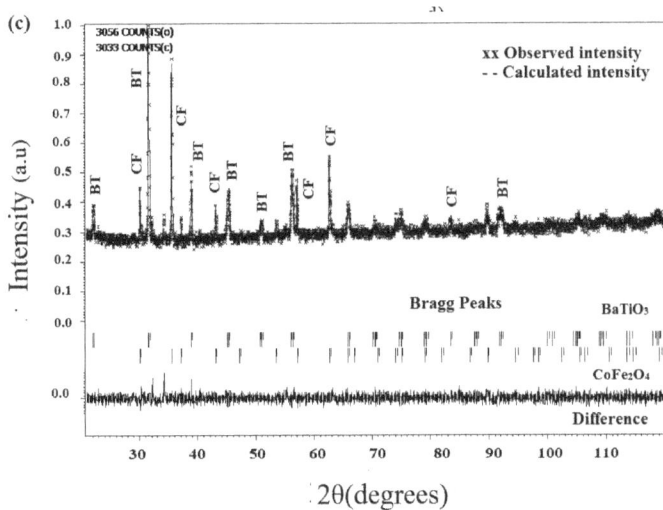

Figure 2.6(c) *Fitted XRD profile of (1-x)BaTiO₃ + xCoFe₂O₄ prepared composite for composition x = 0.6.*

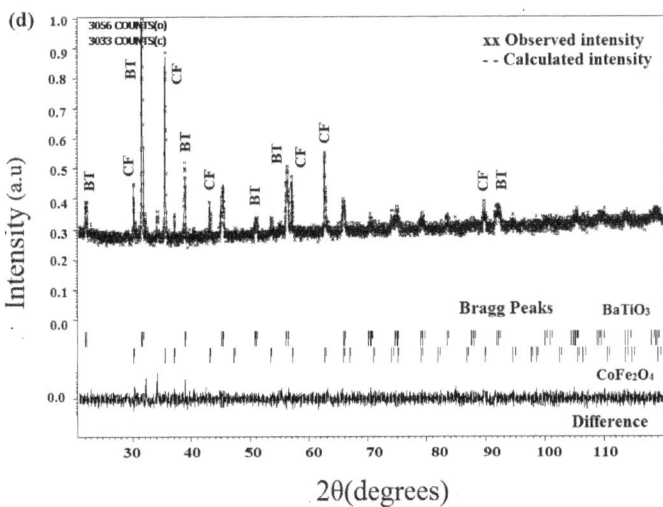

Figure 2.6(d) *Fitted XRD profile of (1-x)BaTiO₃ + xCoFe₂O₄ prepared composite for composition x = 0.8.*

Materials Research Forum LLC
https://doi.org/10.21741/9781644902196

Table 2.3   *Lattice parameters and volume of (1-x) BaTiO₃ + x CoFe₂O₄ composite.*

Composition	BaTiO₃		CoFe₂O₄	Volume (Å³)		c/a
x	a (Å)	c (Å)	a (Å)	Ferroelectric	Ferrite	
0.2	3.9995 (0)	4.0281 (1)	8.3894 (0)	64.43 (0)	590.47 (5)	1.0072
0.4	3.9993 (5)	4.0264 (5)	8.3868 (10)	64.40 (2)	589.93 (8)	1.0068
0.6	4.0003 (5)	4.0237 (6)	8.3875 (9)	64.40 (2)	590.07 (8)	1.0058
0.8	3.9968 (0)	4.0189 (0)	8.3778 (0)	64.20 (0)	588.03 (0)	1.0055

2.2.4   (1-x)BaTiO₃ + xMgFe₂O₄

Solid state reaction method is used to synthesize barium titanate-cobalt ferrite composite for the compositions 0, 0.2, 0.4, 0.6, and 1. The raw XRD profiles are presented in Figure 2.7. The refined profiles are depicted in Figure 2.8 (a) for x = 0, (b) for x = 0.2, (c) for x = 0.4, and (d) for x = 1. The lattice parameters and volume from the Rietveld refinement are given in table 2.4.

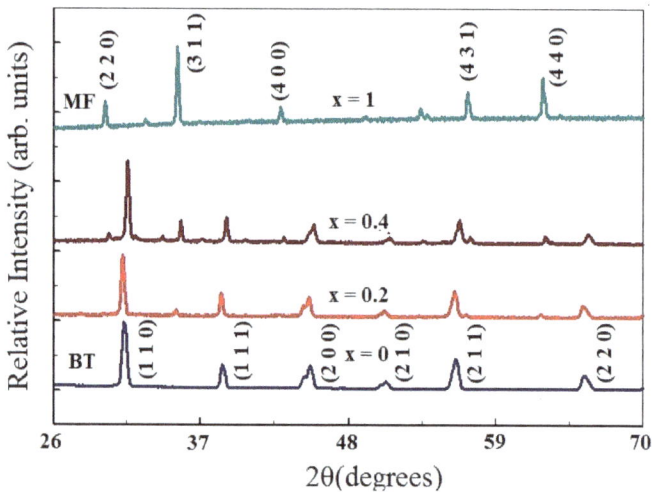

Figure 2.7 *Raw XRD profile of (1-x)BaTiO₃ + xMgFe₂O₄ for x = 0, 0.2, 0.4, and 1.0.*

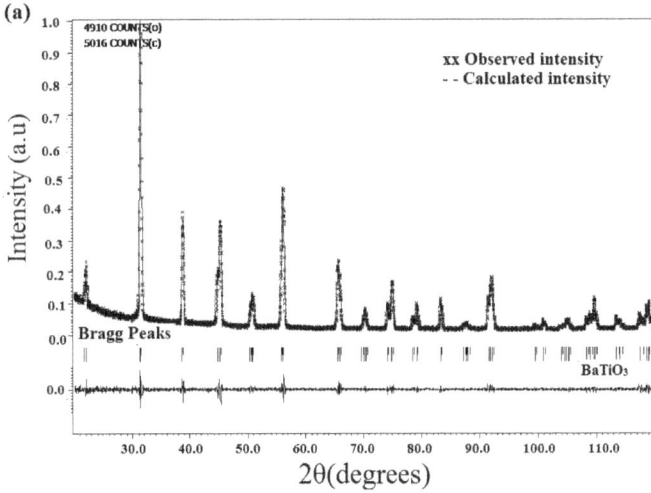

Figure 2.8(a) *Fitted XRD profile of BaTiO₃ in the prepared composite for composition x = 0.*

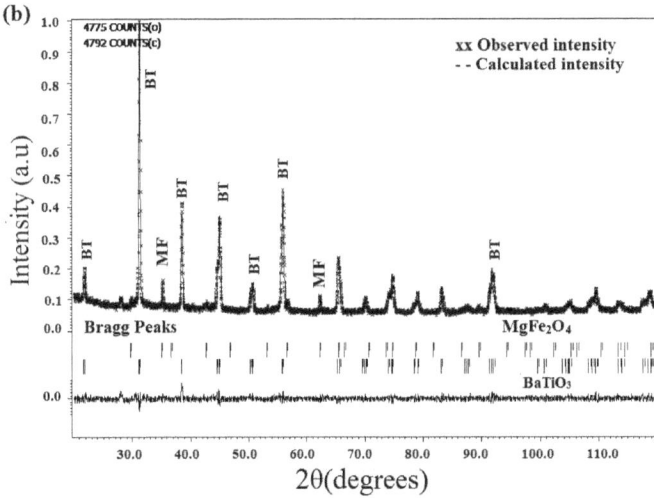

Figure 2.8(b) *Fitted XRD profile of (1-x)BaTiO₃ + xMgFe₂O₄ prepared composite for composition x = 0.2.*

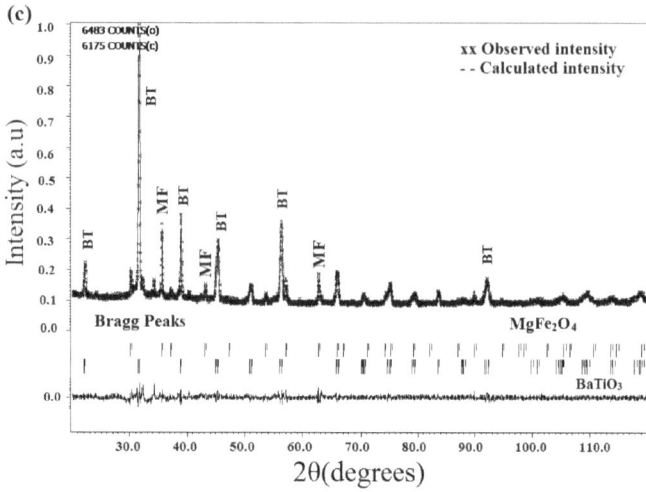

Figure 2.8(c) *Fitted XRD profile of (1-x)BaTiO₃ + xMgFe₂O₄ prepared composite for composition x = 0.4.*

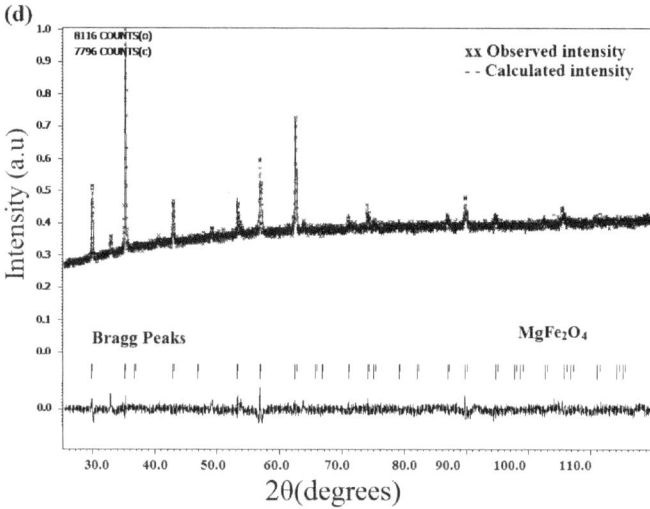

Figure 2.8(d) *Fitted XRD profile of MgFe₂O₄ in the prepared composite for composition x = 1.*

Table 2.4   Lattice parameters and volume of (1-x) BaTiO₃ + x MgFe₂O₄ composite.

Composition	BaTiO₃		MgFe₂O₄	Volume (Å³)		c/a ratio
x	a (Å)	c (Å)	a (Å)	Ferroelectric	Ferrite	
0	4.0022 (2)	4.0413 (2)	-------	64.73 (1)	------	1.0098
0.2	3.9918 (0)	4.0255 (0)	8.3729 (0)	64.14 (0)	586.99 (0)	1.0084
0.4	3.9989 (1)	4.0278 (2)	8.3825 (0)	64.41 (1)	588.99 (2)	1.0072
1.0	------	-------	8.4313 (9)	------	601.02 (04)	

2.3 Morphological studies: SEM images

2.3.1   (1-x)BaTiO₃ + xNiFe₂O₄

The morphology and microstructure of $(1-x)BaTiO_3 + xNiFe_2O_4$ (x=0.2, 0.4, 0.6, and 0.8) have been analyzed from SEM images. The SEM micrographs of $(1-x)BaTiO_3 + xNiFe_2O_4$ corresponding to ×6500 to 7890 magnification are given in figure 2.9 (a) x = 0.2, (b) x=0.4, (c) x=0.6, and x=0.8. The average particle size and grain size are given in Table 2.5.

Figure 2.9 *SEM micrographs of (1-x)BaTiO₃ + xNiFe₂O₄ (a) x = 0.2, (b) x = 0.4, (c) x = 0.6, (d) x = 0.8.*

Table 2.5 *Average and individual particle size for (1-x) BaTiO₃ + x NiFe₂O₄ with compositions (x = 0.2, 0.4, 0.6, 0.8)*

x	Avg. particle size (μm)	Individual particle size BaTiO₃ (nm)	Individual particle size NiFe₂O₄ (nm)
0.2	3.27(12)	25.98(9)	16.24(3)
0.4	3.40 (5)	26.18(8)	21.56(2)
0.6	3.30 (3)	32.29(8)	35.27(3)
0.8	1.74 (1)	23.52(9)	27.66(1)

2.3.2 (1-x)BaTiO₃ + xZnFe₂O₄

The morphology and microstructure of $(1-x)BaTiO_3 + xZnFe_2O_4$ (x=0.2, 0.4, 0.6, and 0.8) have been analyzed from SEM images. The SEM micrographs of $(1-x)BaTiO_3 + xZnFe_2O_4$ corresponding to ×7000 magnification are given in figure 2.10 (a) x = 0.2, (b) x=0.4, (c) x=0.6, and x=0.8. The average particle size and grain size are given in Table 2.6.

Figure 2.10 *SEM micrographs of (1-x)BaTiO₃ + xZnFe₂O₄ (a) x = 0.2, (b) x = 0.4, (c) x = 0.6, (d) x = 0.8.*

Table 2.6 *Average and individual particle size for (1-x) BaTiO$_3$ + x ZnFe$_2$O$_4$ (x = 0.2, 0.4, 0.6, 0.8)*

x	Avg. particle size (μm)	Individual particle size BaTiO$_3$ (nm)	Individual particle size ZnFe$_2$O$_4$ (nm)
0.2	1.021 (3)	22.33 (1.6)	24.86 (7.6)
0.4	1.209 (1)	24.15 (1.7)	22.29 (3.1)
0.6	1.574 (2)	24.38 (1.8)	24.42 (4.0)
0.8	1.575 (1)	24.43 (3.3)	28.25 (0.36)

2.3.3 (1-x)BaTiO$_3$ + xCoFe$_2$O$_4$

The morphology and microstructure of (1-x)BaTiO$_3$ + xCoFe$_2$O$_4$ (x=0.2, 0.4, 0.6, and 0.8) have been analyzed from SEM images. The SEM micrographs of (1-x)BaTiO$_3$ + xCoFe$_2$O$_4$ corresponding to ×25000 magnification are given in figure 2.11 (a) x = 0.2, (b) x=0.4, (c) x=0.6, and (d) x=0.8. The average particle size and grain size are given in Table 2.7.

Figure 2.11 *SEM micrographs of (1-x)BaTiO$_3$ + xCoFe$_2$O$_4$ (a) x = 0.2, (b) x = 0.4, (c) x = 0.6, (d) x = 0.8.*

Table 2.7   *Average and individual particle size for (1-x)BaTiO$_3$ + xCoFe$_2$O$_4$ composite.*

x	Avg. particle size (μm)	Individual particle size (nm) BaTiO$_3$	Individual particle size (nm) CoFe$_2$O$_4$
0.2	1.68 (9)	29.38 (3)	32.29 (5)
0.4	1.43 (9)	34.98 (7)	34.78 (3)
0.6	1.82 (7)	29.38 (3)	39.42 (4)
0.8	1.48 (8)	31.03 (3)	36.97 (2)

2.3.4 (1-x)BaTiO$_3$ + xMgFe$_2$O$_4$

The morphology and microstructure of (1-x)BaTiO$_3$ + xMgFe$_2$O$_4$ (x=0, 0.2, 0.4, and 1.0) have been analyzed from SEM images. The SEM micrographs of (1-x)BaTiO$_3$ + xMgFe$_2$O$_4$ corresponding to ×5000 to x10000 magnification are given in figure 2.12 (a) x = 0, (b) x=0.2, (c) x=0.4, and (d) x=1.0. The values of the particle size and the grain size are given in table 2.8.

Figure 2.12 *SEM micrographs of (1-x)BaTiO$_3$ + xMgFe$_2$O$_4$ (a) x = 0, (b) x = 0.2, (c) x = 0.4, (d) x = 1.0.*

Table 2.8 *Average and individual particle size for (1-x) BaTiO₃ + x MgFe₂O₄ (x = 0. 0.2, 0.4, 1)*

x	Avg. particle size (μm)	Individual particle size BaTiO₃ (nm)	Individual particle size MgFe₂O₄ (nm)
0	1.384	17.46 (2)	--
0.2	1.119	26.01 (4)	36.76 (9)
0.4	0.966	20.06 (2)	28.93 (5)
1	1.126	--	35.86 (1)

2.4 Elemental analysis - EDS

2.4.1 (1-x)BaTiO₃ + xNiFe₂O₄

The elemental compositions of the magneto-electric ceramic composite $(1-x)BaTiO_3 + xNiFe_2O_4$ (x=0.2, 0.4, 0.6, and 0.8) have been analyzed qualitatively and quantitatively using energy dispersive X-ray spectroscopy (EDS). The EDS spectra of $(1-x)BaTiO_3 + xNiFe_2O_4$ ceramic composite for the compositions x=0.2, 0.4, 0.6, and 0.8 are shown in figure 2.13 (a) for x = 0.2, (b) for x = 0.4, (c) for x = 0.6 and (d) x = 0.8. Table 2.9 gives the atomic percentages and mass percentages of $(1-x)BaTiO_3 + xNiFe_2O_4$.

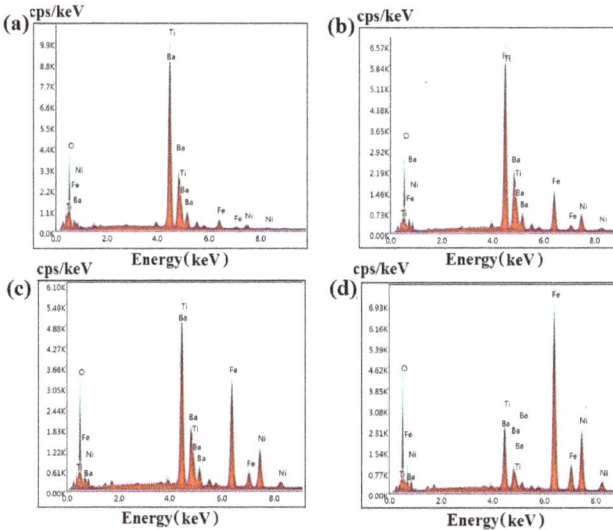

Figure 2.13 *EDS spectra of (1-x)BaTiO₃ + xNiFe₂O₄ (a) x = 0.2, (b) x = 0.4, (c) x = 0.6, and (d) x = 0.8.*

Table 2.9 *Elemental compositions of (1-x) BaTiO₃ + x NiFe₂O₄ from EDS analysis*

Composition (x)	Atomic percentage (%)					Weight percentage (%)				
	Ba	Ti	Ni	Fe	O	Ba	Ti	Ni	Fe	O
x = 0.2	12.57	13.31	0.81	1.8	65.21	46.72	17.26	1.29	2.72	28.24
x = 0.4	13.10	12.5	3.63	7.88	58.22	44.13	14.68	5.23	10.79	22.85
x = 0.6	10.13	9.29	3.22	15.85	58.23	34.48	11.02	9.47	21.94	23.09
x = 0.8	4.45	3.12	11.41	27.59	53.44	15.96	3.92	17.52	40.27	22.35

2.4.2 (1-x)BaTiO₃ + xZnFe₂O₄

The elemental compositions of the magneto-electric ceramic composite $(1-x)BaTiO_3 + xZnFe_2O_4$ (x=0.2, 0.4, 0.6, and 0.8) have been analyzed qualitatively and quantitatively using energy dispersive X-ray spectroscopy (EDS). The EDS spectra of $(1-x)BaTiO_3 + xZnFe_2O_4$ ceramic composite for the compositions x=0.2, 0.4, 0.6, and 0.8 are shown in figure 2.14 (a) for x = 0.2, (b) for x = 0.4, (c) for x = 0.6 and (d) x = 0.8. The atomic percentages and mass percentages of $(1-x)BaTiO_3 + xZnFe_2O_4$ are given in table 2.10.

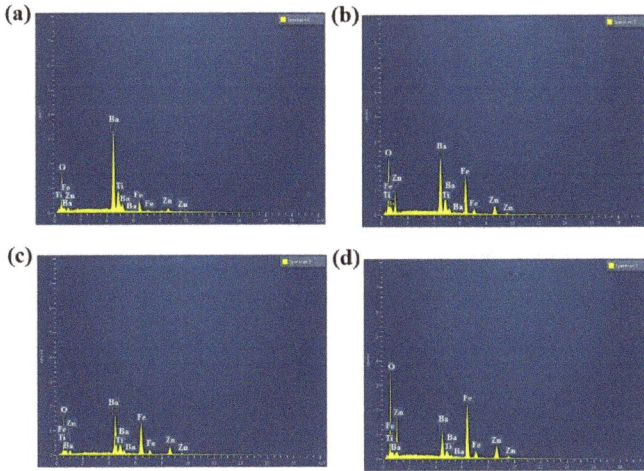

Figure 2.14 EDS spectra of (1-x)BaTiO₃ + xZnFe₂O₄ (a) x = 0.2, (b) x = 0.4, (c) x = 0.6, and (d) x = 0.8.

Table 2.10 Elemental compositions of (1-x) BaTiO₃ + x ZnFe₂O₄ from EDS analysis

Composition(x)	Atomic percentage (%)					Weight percentage (%)				
	Ba	Ti	Zn	Fe	O	Ba	Ti	Zn	Fe	O
x = 0.2	15.91	15.4	3.81	6.95	58.65	47.43	16.76	5.66	8.82	21.33
x = 0.4	7.77	8.66	8.61	18.77	56.2	26.74	10.39	14.09	26.26	22.53
x = 0.6	8.04	7.63	10.69	21.65	51.99	26.23	8.68	16.6	28.73	19.76
x = 0.8	2.77	3.29	10.98	22.2	60.76	10.97	4.54	20.7	35.75	28.03

2.4.3 **(1-x)BaTiO₃ + xCoFe₂O₄**

The elemental compositions of the magneto-electric ceramic composite $(1-x)BaTiO_3$ + $xCoFe_2O_4$ (x=0.2, 0.4, 0.6, and 0.8) have been analyzed qualitatively and quantitatively using energy dispersive X-ray spectroscopy (EDS). The EDS spectra of $(1-x)BaTiO_3$ + $xCoFe_2O_4$ ceramic composite for the compositions x=0.2, 0.4, 0.6, and 0.8 are shown in figure 2.15 (a) for x = 0.2, (b) for x = 0.4, (c) for x = 0.6 and (d) x = 0.8. The atomic percentages and mass percentages of $(1-x)BaTiO_3$ + $xCoFe_2O_4$ are given in table 2.11.

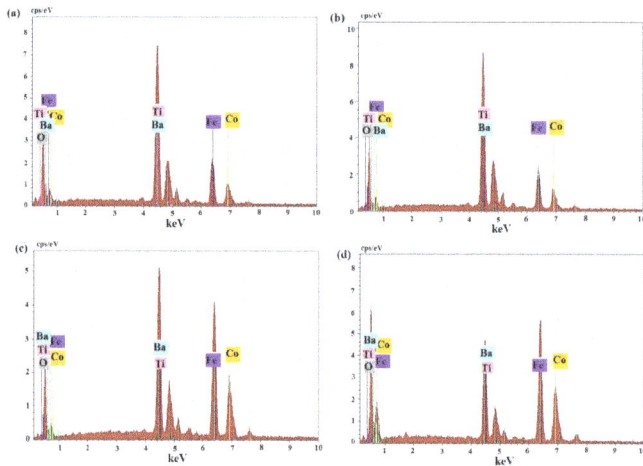

Figure 2.15 *EDS spectra of (1-x)BaTiO₃ + xCoFe₂O₄ (a) x = 0.2, (b) x = 0.4, (c) x = 0.6, and (d) x = 0.8.*

Table 2.11 *Elemental compositions of (1-x)BaTiO₃ + xCoFe₂O₄ from EDS analysis*

Composition x	Atomic percentage (%)					Weight percentage (%)				
	Ba	Ti	Co	Fe	O	Ba	Ti	Co	Fe	O
x = 0.2	9.20	7.84	4.09	8.25	70.62	36.41	10.81	6.95	13.27	32.55
x = 0.4	11.46	8.2	4.74	9.04	66.56	41.24	10.31	7.31	13.23	27.91
x = 0.6	8.51	6.65	9.10	19.77	55.97	29.04	7.92	13.34	27.45	22.26
x = 0.8	4.07	3.13	7.37	14.07	71.35	18.21	4.89	14.14	25.59	37.17

2.4.4   (1-x)BaTiO₃ + xMgFe₂O₄

The elemental compositions of the magneto-electric ceramic composite $(1-x)BaTiO_3$ + $xCoFe_2O_4$ (x=0.2, 0.4) have been analyzed qualitatively and quantitatively using energy dispersive X-ray spectroscopy (EDS). The EDS spectra of $(1-x)BaTiO_3$ + $xMgFe_2O_4$ ceramic composite for the compositions x=0.2 and 0.4are shown in figure 2.16 (a) for x = 0.2, (b) for x = 0.4. The atomic percentages and mass percentages of $(1-x)BaTiO_3$ + $xCoFe_2O_4$ are given in table 2.12.

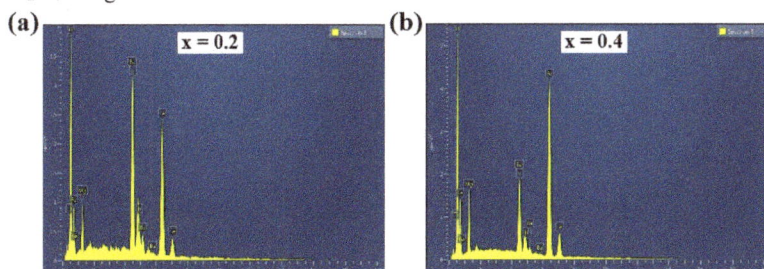

Figure 2.16 *EDS spectra of (1-x)BaTiO₃ + xMgFe₂O₄ (a) x = 0.2, (b) x = 0.4.*

Table 2.12   *Elemental compositions of (1-x)BaTiO₃ + xMgFe₂O₄ from EDS analysis*

Composition x	Atomic percentage (%)					Weight percentage (%)				
	Ba	**Ti**	**Mg**	**Fe**	**O**	**Ba**	**Ti**	**Mg**	**Fe**	**O**
x = 0.2	8.53	7.94	8.6	20.96	53.97	30.89	10.02	5.52	30.84	22.71
x = 0.4	3.9	3.42	10.06	30.91	51.71	15.33	4.66	7.02	49.33	23.63

2.5   Optical characterization – UV -visible absorption spectra

2.5.1   (1-x)BaTiO₃ + xNiFe₂O₄

The optical band gap value has been evaluated using UV-vis absorption data obtained in the wavelength range of 200 nm to 2000 nm. Figure 2.17 shows the UV-visible absorption spectra of $(1-x)BaTiO_3 + xNiFe_2O_4$. Since, $NiFe_2O_4$ is an direct/indirect bandgap material [Hollinsworth et al., 2014]. The Tauc plot for $(1-x)BaTiO_3 + xNiFe_2O_4$ magneto-electric ceramic composite is shown in figure 2.18 and the values are given in table 2.13.

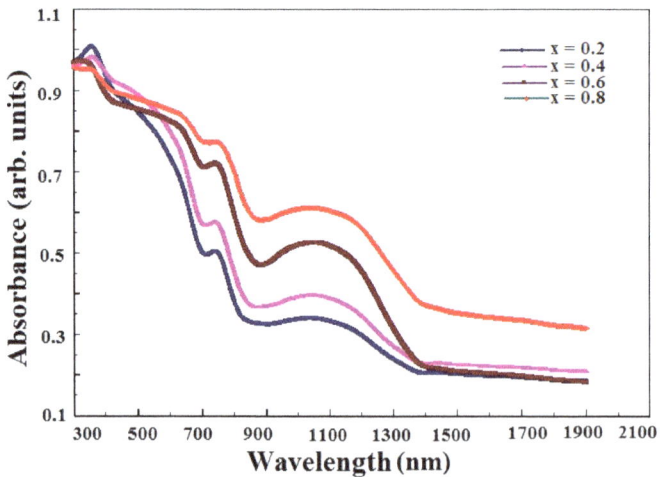

Figure 2.17 *UV-Visible absorption spectra of (1-x)BaTiO₃ + xNiFe₂O₄.*

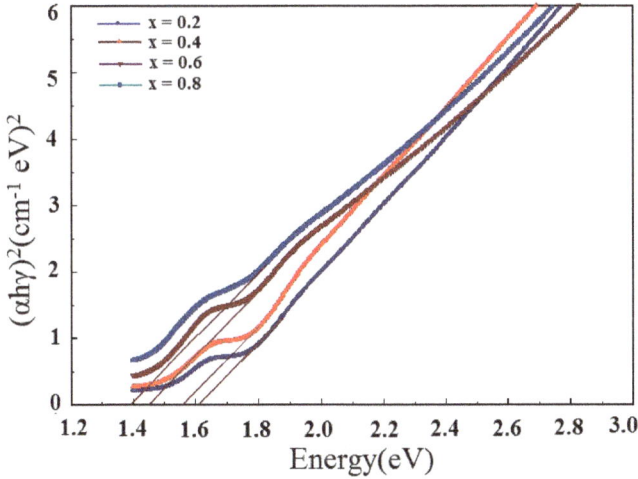

Figure 2.18 *Tauc Plot of the (1-x)BaTiO₃ + xNiFe₂O₄ for x = 0.2, 0.4, 0.6, and 0.8.*

Table 2.13 *The optical band gap for the prepared composite (1-x) BaTiO₃ + x NiFe₂O₄*

Compositions	Optical band gap (eV)
x	
x = 0.2	1.612
x = 0.4	1.560
x = 0.6	1.454
x = 0.8	1.393

2.5.2 (1-x)BaTiO₃ + xZnFe₂O₄

From the UV-Visible absorption data, the optical band gap value has been evaluated in the wavelength range of 200 nm to 2000 nm. Figure 2.19 shows the UV-visible absorption spectra of $(1-x)BaTiO_3 + xZnFe_2O_4$. The Tauc plot of $(1-x)BaTiO_3 + xZnFe_2O_4$ magneto-electric ceramic composite has been drawn using the Tauc's relation [Wood and Tauc, 1972] and is given in figure 2.20. The optical band gap values for $(1-x)BaTiO_3 + xZnFe_2O_4$ for compositions x = 0.2, 0.4, 0.6, and 0.8 are given in table 2.14.

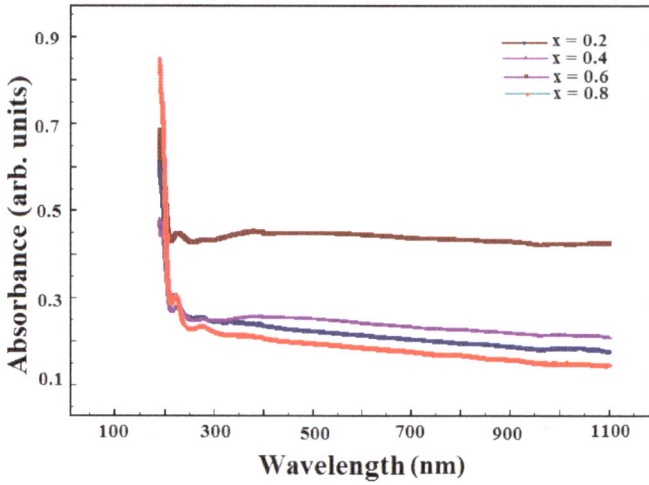

Figure 2.19 *UV-visible absorption spectra of (1-x)BaTiO₃ + xZnFe₂O₄.*

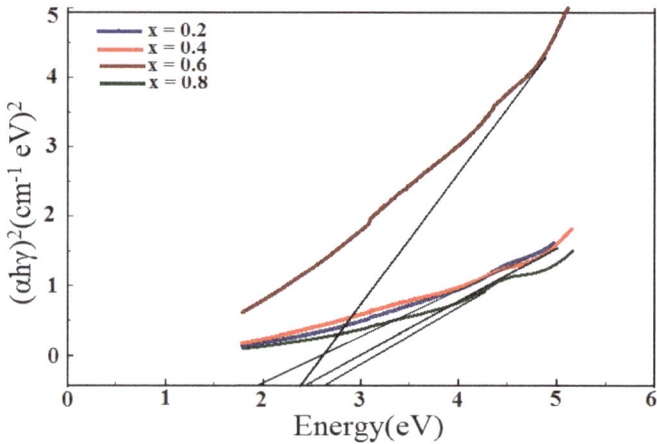

Figure 2.20 *Tauc Plot of the (1-x)BaTiO₃ + xZnFe₂O₄ for x = 0.2, 0.4, 0.6, and 0.8.*

Table 2.14 *Optical band gap for the prepared composite ((1-x) BaTiO₃ + x ZnFe₂O₄)*

Compositions	Optical band gap (eV)
x = 0.2	1.95
x = 0.4	2.44
x = 0.6	2.38
x = 0.8	2.64

2.5.3   (1-x)BaTiO₃ + xCoFe₂O₄

From the UV-Visible absorption data, the optical band gap value has been evaluated in the wavelength range of 200 nm to 2000 nm. Figure 2.21 shows the UV-visible absorption spectra of $(1-x)BaTiO_3 + xCoFe_2O_4$. The Tauc plot of $(1-x)BaTiO_3 + xCoFe_2O_4$ magneto-electric ceramic composite has been drawn using the Tauc's relation [Wood and Tauc, 1972] and is given in figure 2.22. The optical band gap values for $(1-x)BaTiO_3 + xZnFe_2O_4$ for compositions x = 0.2, 0.4, 0.6, and 0.8 are given in table 2.15.

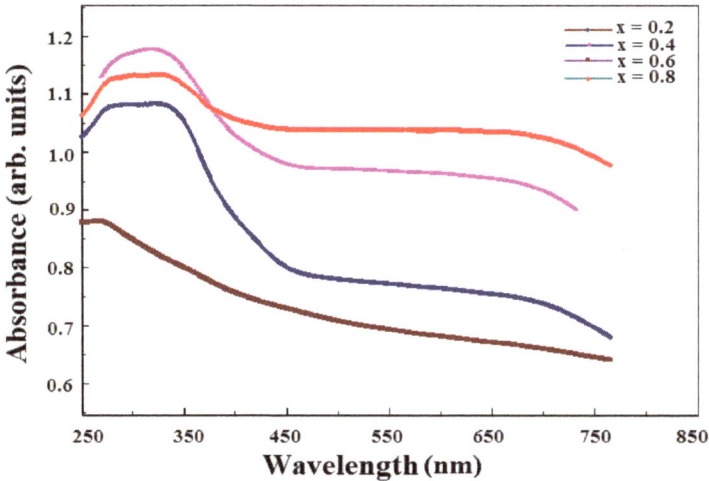

Figure 2.21 *UV-visible absorption spectra of $(1-x)BaTiO_3 + xCoFe_2O_4$.*

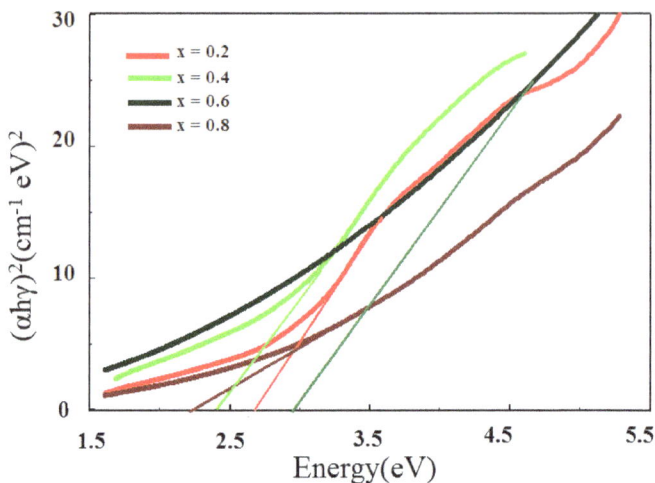

Figure 2.22 *Tauc Plot of the (1-x)BaTiO₃ + xCoFe₂O₄ for x = 0.2, 0.4, 0.6, and 0.8.*

Table 2.15 *The optical band gap for the prepared composite ((1-x)BaTiO₃ + xCoFe₂O₄)*

Compositions	Optical band gap (eV)
x = 0.2	2.67
x = 0.4	2.41
x = 0.6	2.94
x = 0.8	2.22

2.5.4   (1-x)BaTiO₃ + xMgFe₂O₄

From the UV-Visible absorption data, the optical band gap value has been evaluated in the wavelength range of 200 nm to 2000 nm. Figure 2.23 shows the UV-visible absorption spectra of $(1-x)BaTiO_3 + xCoFe_2O_4$. The Tauc plot of $(1-x)BaTiO_3 + xMgFe_2O_4$ magneto-electric ceramic composite has been drawn using the Tauc's relation [Wood and Tauc, 1972] and is given in figure 2.24. The optical band gap values for $(1-x)BaTiO_3 + xZnFe_2O_4$ for compositions x = 0.2, 0.4, 0.6, and 0.8 are given in table 2.16.

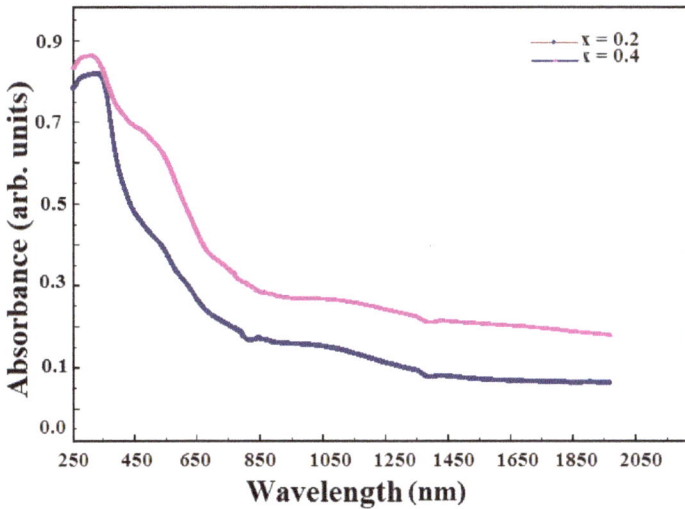

Figure 2.23 *UV-visible absorption spectra of (1-x)BaTiO₃ + xCoFe₂O₄*

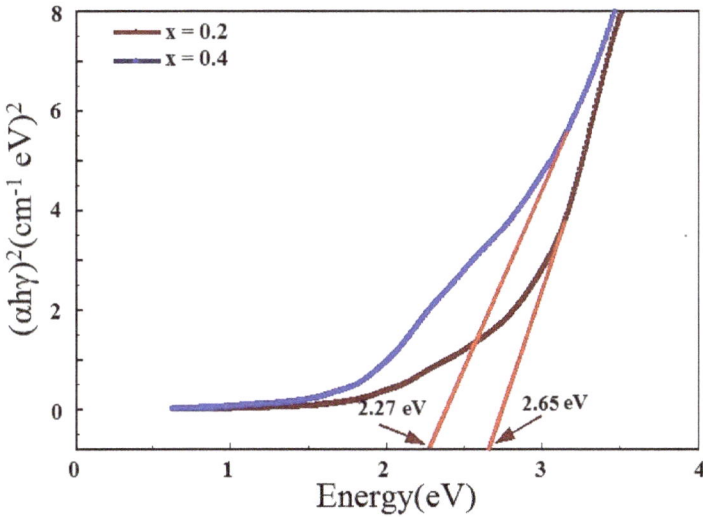

Figure 2.24 *Tauc Plot of the (1-x)BaTiO₃ + xCoFe₂O₄ for x = 0.2 and 0.4.*

Table 2.16 *The optical band gap for the prepared composite ((1-x)BaTiO$_3$ + xMgFe$_2$O$_4$)*

Compositions	Optical band gap (eV)
x = 0.2	2.65
x = 0.4	2.27

2.6   Electrical characterization – Dielectric and P-E studies.

2.6.1   (1-x)BaTiO$_3$ + xNiFe$_2$O$_4$

The electrical characterization is investigated through the dielectric and P-E studies. The dielectric and P-E characterization of (1-x)BaTiO$_3$ + xNiFe$_2$O$_4$ ceramic composite is shown in Figures 2.25 and 2.26. Figure 2.25 (a) depicts the capacitance vs frequency graph and Figure 2.25 (b) shows the dielectric loss vs frequency graph. Figure 2.26 represents P-E hysteresis of the prepared composite. The quantitative details of the P-E characterization are given in table 2.17.

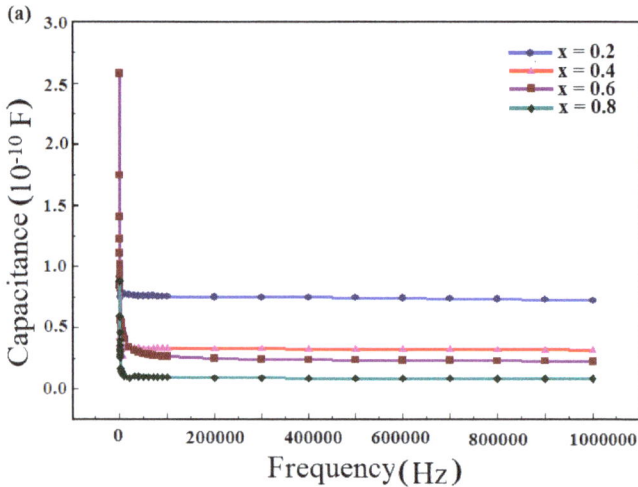

Figure 2.25 (a) *The capacitance-frequency graph of (1-x)BaTiO$_3$ + xNiFe$_2$O$_4$ ceramic composite for x = 0.2, 0.4, 0.6, and 0.8.*

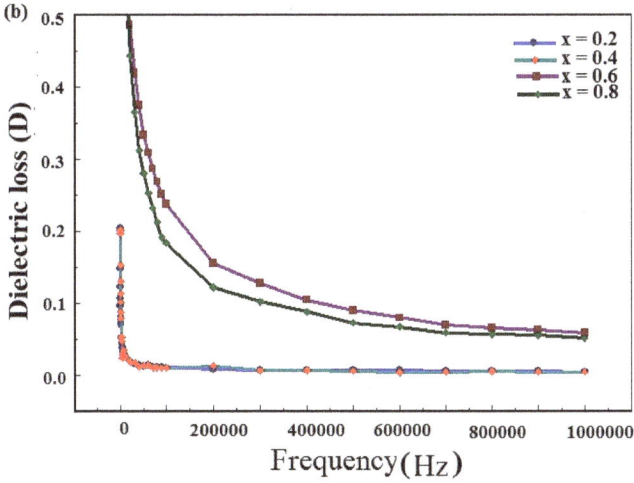

Figure 2.25 (b) *The dielectric loss-frequency graph of (1-x)BaTiO₃ + xNiFe₂O₄ ceramic composite for x = 0.2, 0.4, 0.6, and 0.8.*

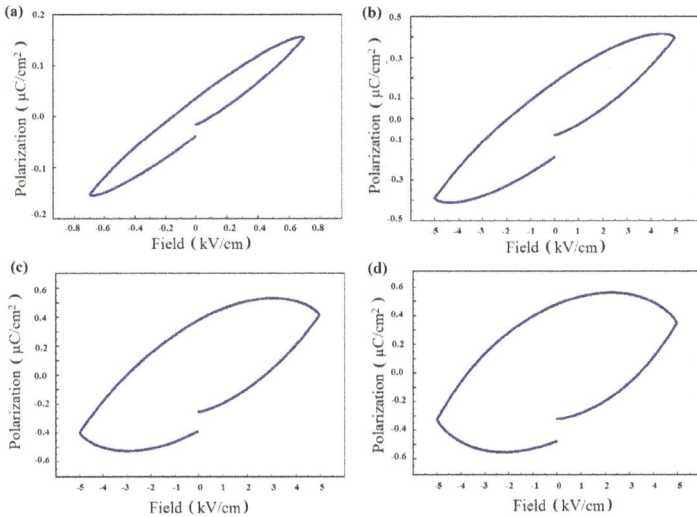

Figure 2.26 *P-E hysteresis of (1-x) BaTiO₃ + x NiFe₂O₄ (a) x = 0.2, (b) x = 0.4, (c) x = 0.6, and (d) x = 0.8.*

Table 2.17 *The electric parameters of (1-x) BaTiO₃ + x NiFe₂O₄ composite from the study of P-E (electric) hysteresis*

Composition	P_{max} $(\mu C/cm^2)$	P_r $(\mu C/cm^2)$	$-P_r$ $(\mu C/cm^2)$
x = 0.2	0.154	0.035	0.039
x = 0.4	0.393	0.175	0.192
x = 0.6	0.412	0.0	0.393
x = 0.8	0.342	0.475	0.481

2.6.2 (1-x)BaTiO₃ + xZnFe₂O₄

The electrical characterization is investigated through the dielectric and P-E studies. The dielectric and P-E characterization of (1-x)BaTiO₃ + xZnFe₂O₄ ceramic composite is shown in Figures 2.27 and 2.28. Figure 2.27 (a) depicts the capacitance vs frequency graph and Figure 2.27 (b) shows the dielectric loss vs frequency graph. Figure 2.28 represents P-E hysteresis of the prepared composite. The quantitative details of the P-E characterization are given in table 2.18.

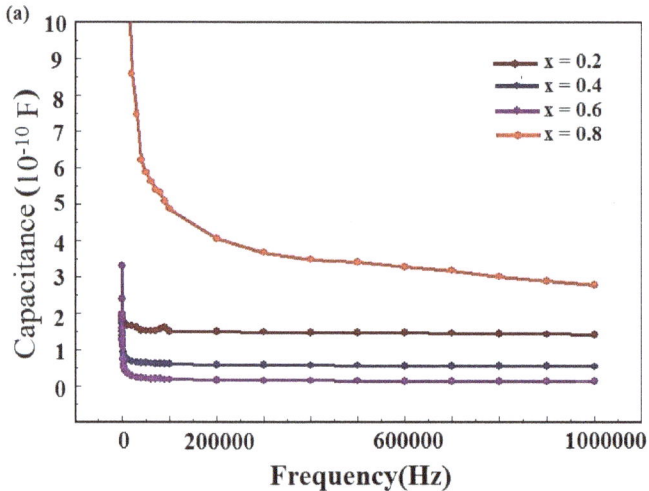

Figure 2.27 (a) *The capacitance-frequency graph of (1-x)BaTiO₃ + xZnFe₂O₄ ceramic composite for x = 0.2, 0.4, 0.6, and 0.8.*

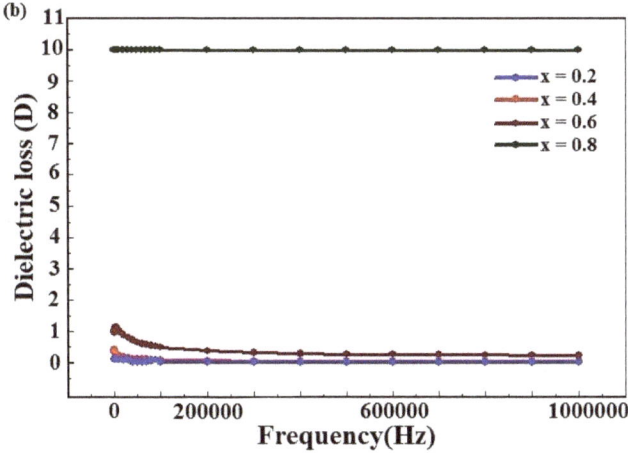

Figure 2.27 (b) *The dielectric loss-frequency graph of (1-x)BaTiO₃ + xZnFe₂O₄ ceramic composite for x = 0.2, 0.4, 0.6, and 0.8.*

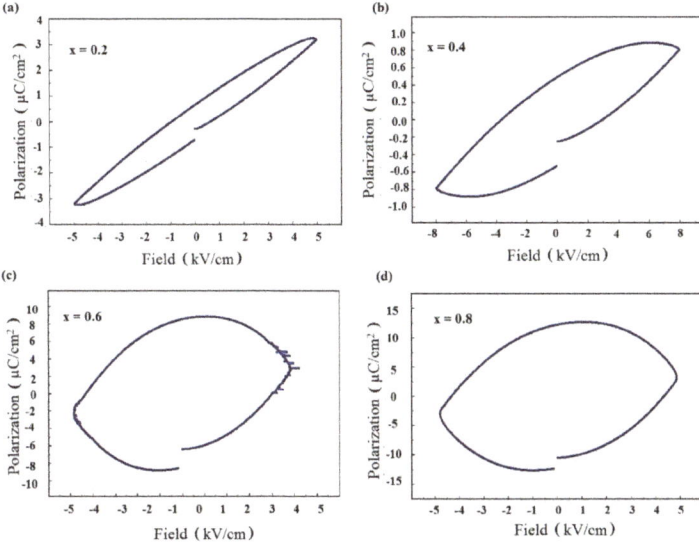

Figure 2.28 *P-E hysteresis of (1-x) BaTiO₃ + x ZnFe₂O₄ (a) x = 0.2, (b) x = 0.4, (c) x = 0.6, and (d) x = 0.8.*

Table 2.18 *The electric parameters of (1-x) BaTiO₃ + x ZnFe₂O₄ composite from the study of P-E (electric) hysteresis*

Composition	P_{max} ($\mu C/cm^2$)	P_r ($\mu C/cm^2$)	$-P_r$ ($\mu C/cm^2$)
x = 0.2	0.321	0.0647	-0.0713
x = 0.4	0.804	0.489	-0.531
x = 0.6	2.92	8.50	-8.59
x = 0.8	3.29	12.2	-12.4

2.6.3 (1-x)BaTiO₃ + xCoFe₂O₄

The electrical characterization is investigated through the dielectric and P-E studies. The dielectric and P-E characterization of (1-x)BaTiO₃ + xCoFe₂O₄ ceramic composite is shown in Figures 2.29 and 2.30. Figure 2.29 (a) depicts the capacitance vs frequency graph and Figure 2.29 (b) shows the dielectric loss vs frequency graph. Figure 2.30 represents P-E hysteresis of the prepared composite. The quantitative details of the P-E characterization are given in Table 2.19.

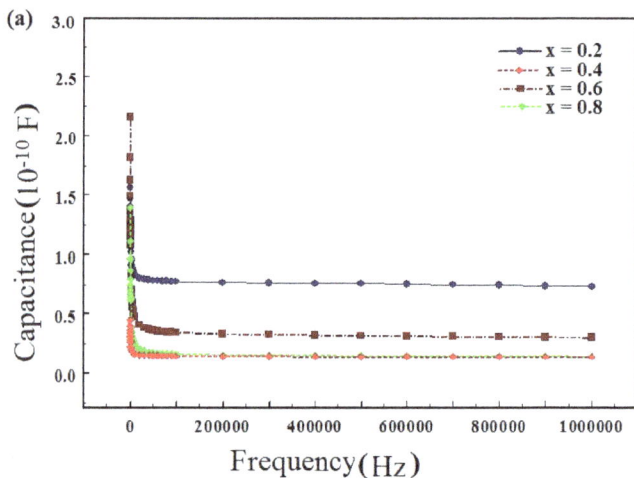

Figure 2.29 (a) *The capacitance-frequency graph of (1-x)BaTiO₃ + xCoFe₂O₄ ceramic composite for x = 0.2, 0.4, 0.6, and 0.8.*

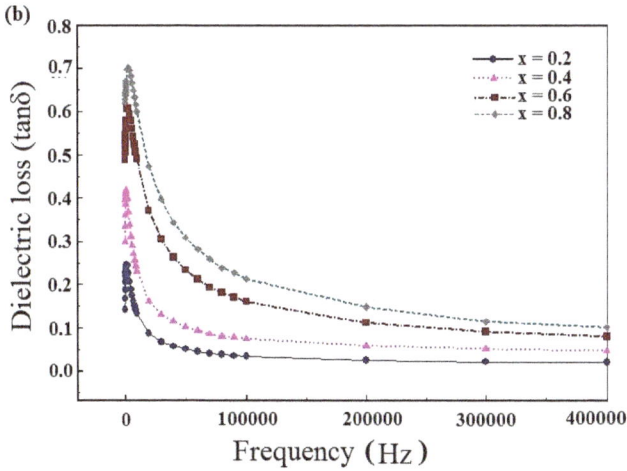

Figure 2.29 (b) *The dielectric loss-frequency graph of (1-x)BaTiO₃ + xCoFe₂O₄ ceramic composite for x = 0.2, 0.4, 0.6, and 0.8.*

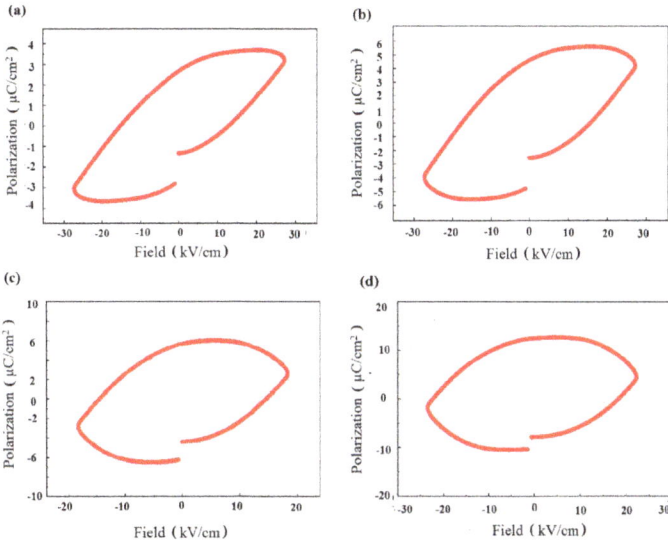

Figure 2.30 *P-E hysteresis of (1-x) BaTiO₃ + x CoFe₂O₄ (a) x = 0.2, (b) x = 0.4, (c) x = 0.6, and (d) x = 0.8.*

Table 2.19 *The electric parameters of (1-x) BaTiO₃ + x CoFe₂O₄ composite from the study of P-E (electric) hysteresis*

Composition x	P_{max} ($\mu C/cm^2$)	P_r ($\mu C/cm^2$)	$-P_r$ ($\mu C/cm^2$)
x = 0.2	3.21	2.63	2.84
x = 0.4	4.18	4.57	4.80
x = 0.6	4.28	8.94	9.10
x = 0.8	4.32	14.2	14.3

2.6.4   (1-x)BaTiO₃ + xMgFe₂O₄

The electrical characterization is investigated through the dielectric and P-E studies. The dielectric characterization of (1-x)BaTiO₃ + xMgFe₂O₄ ceramic composite is shown in Figure 2.31. Figure 2.31 (a) depicts the capacitance vs frequency graph and Figure 2.31(b) shows the dielectric loss vs frequency graph.

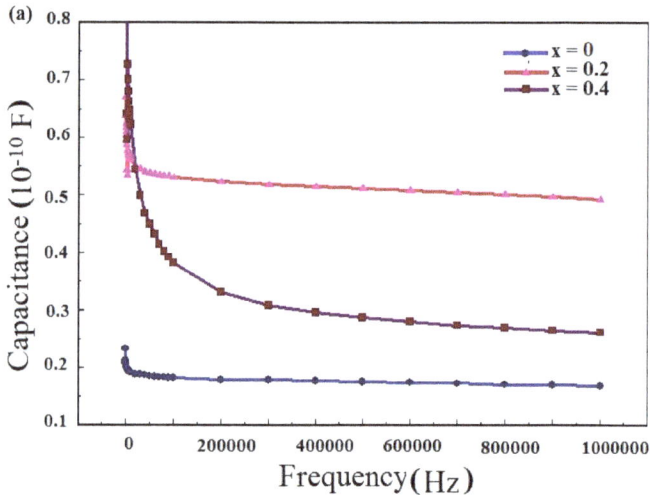

Figure 2.31 (a) *The capacitance-frequency graph of (1-x)BaTiO₃ + xMgFe₂O₄ ceramic composite for x = 0.2, 0.4, 0.6, and 0.8.*

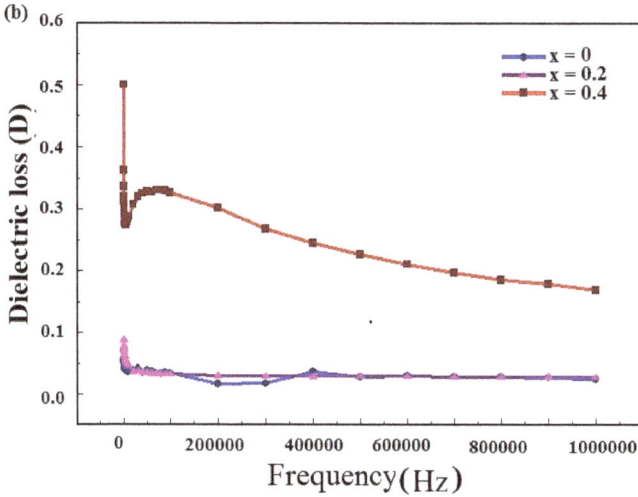

Figure 2.31 (b) *The dielectric loss-frequency graph of (1-x)BaTiO₃ + xMgFe₂O₄ ceramic composite for x = 0.2, 0.4, 0.6, and 0.8.*

2.7 Magnetic characterization – M-H studies

2.7.1 (1-x)BaTiO₃ + xNiFe₂O₄

The magnetic characterization of $(1-x)BaTiO_3 + xNiFe_2O_4$ ceramic composite is investigated through the M-H studies. The M-H characterization of the ceramic composite is shown in **figures 2.32 (a) and (b)**. The values of saturation magnetization, coercivity and retentivity are given in **table 2.20**.

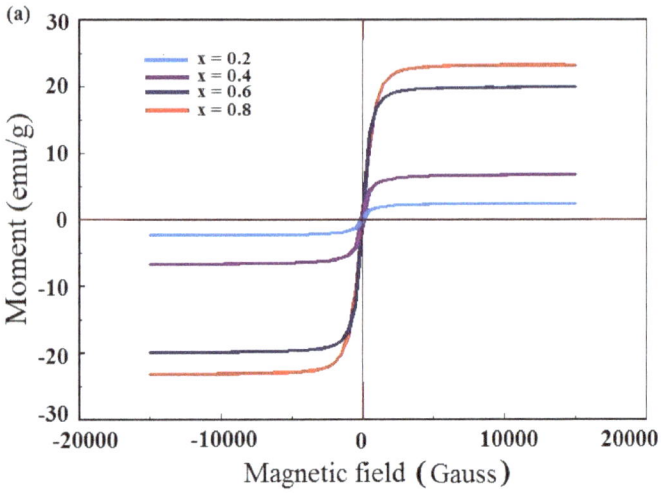

Figure 2.32 (a) *The magnetic characterization (M-H) of (1-x)BaTiO₃ + xNiFe₂O₄ for x = 0.2, 0.4, 0.6 and 0.8.*

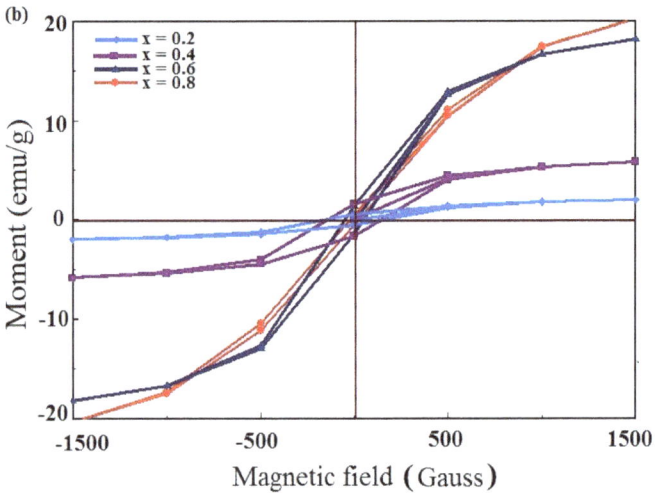

Figure 2.32 (b) *Expanded M-H graph of (1-x)BaTiO₃ + xNiFe₂O₄ for x = 0.2, 0.4, 0.6 and 0.8.*

Table 2.20 *The magnetic parameters of $(1-x)BaTiO_3 + xNiFe_2O_4$ composite from the study of magnetic hysteresis.*

Composition x	Coercivity H_{ci} (G)	Magnetization M_s (emu/g)	Retentivity $M_r \times 10^{-3}$ (emu)
x = 0.2	157.52	2.423	27.68
x = 0.4	146.22	6.712	41.13
x = 0.6	51.34	19.941	31.79
x = 0.8	23.87	23.282	16.18

2.7.2   (1-x)BaTiO₃ + xZnFe₂O₄

The magnetic characterization of $(1-x)BaTiO_3 + xZnFe_2O_4$ ceramic composite is investigated through the M-H studies. The M-H characterization of the ceramic composite is shown in figures 2.33 (a) and (b). The values of saturation magnetization, coercivity and retentivity are given in Table 2.21

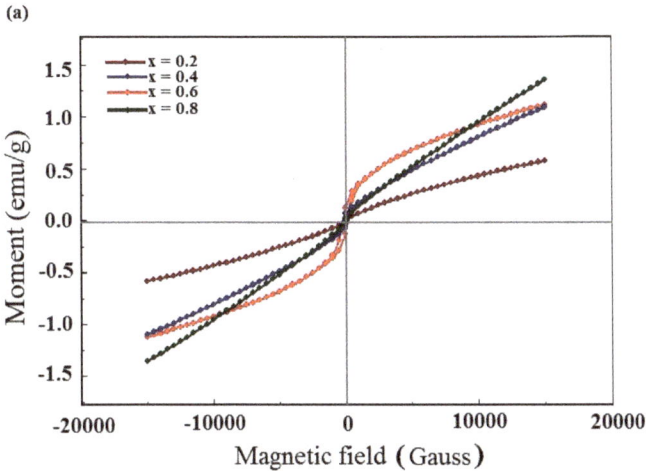

Figure 2.33 (a) *The magnetic characterization (M-H) of $(1-x)BaTiO_3 + xZnFe_2O_4$ for x = 0.2, 0.4, 0.6 and 0.8.*

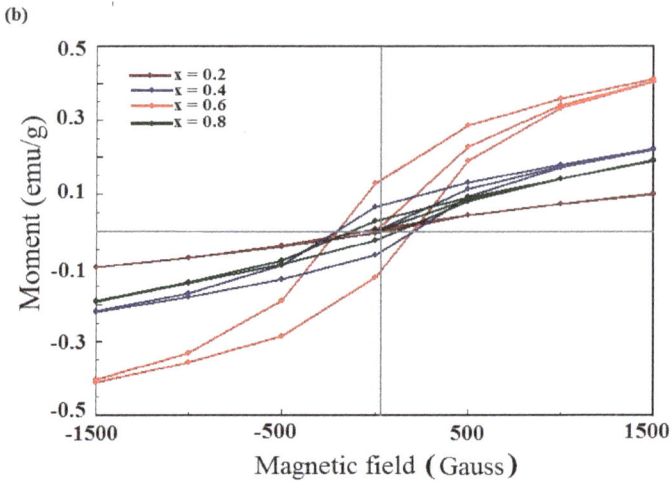

Figure 2.33 (b) *Expanded M-H of (1-x)BaTiO₃ + xZnFe₂O₄ for x = 0.2, 0.4, 0.6 and 0.8.*

Table 2.21 *The magnetic parameters of (1-x) BaTiO₃ + x ZnFe₂O₄ composite from the study of magnetic hysteresis*

Composition x	Coercivity H_{ci} (Oe)	Magnetization M_s (emu/g)	Retentivity M_r x 10^{-3} (emu)
x = 0.2	35.309	0.583	0.165
x = 0.4	200.69	1.119	5.566
x = 0.6	207.35	1.097	3.200
x = 0.8	121.81	1.359	1.554

2.7.3 (1-x)BaTiO₃ + xCoFe₂O₄

The magnetic characterization of (1-x)BaTiO₃ + xCoFe₂O₄ ceramic composite is investigated through the M-H studies. The M-H characterization of the ceramic composite is shown in Figure 2.34. The values of saturation magnetization, coercivity and retentivity are given in table 2.22.

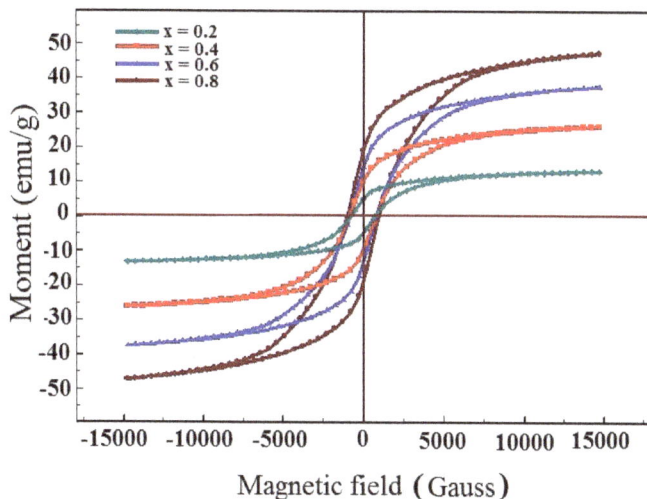

Figure 2.34 *The magnetic characterization (M-H) of (1-x)BaTiO₃ + xCoFe₂O₄ for x = 0.2, 0.4, 0.6 and 0.8.*

Table 2.22 *The saturation magnetization and coercivity of the composite for x = 0.2, 0.4, 0.6, 0.8*

Composition x	M_s (emu/g)	H_c (G)	Retentivity M_r x 10^{-3} (emu)
x = 0.2	13.182	746.38	6.498
x = 0.4	26.156	833.58	10.631
x = 0.6	37.543	816.16	14.765
x = 0.8	47.260	957.78	20.277

2.7.4 (1-x)BaTiO₃ + xMgFe₂O₄

The magnetic characterization of $(1-x)BaTiO_3 + xMgFe_2O_4$ ceramic composite is investigated through the M-H studies. The M-H characterization of the ceramic composite is shown in Figures 2.35 (a) and (b). The values of saturation magnetization, coercivity and retentivity are given in table 2.23.

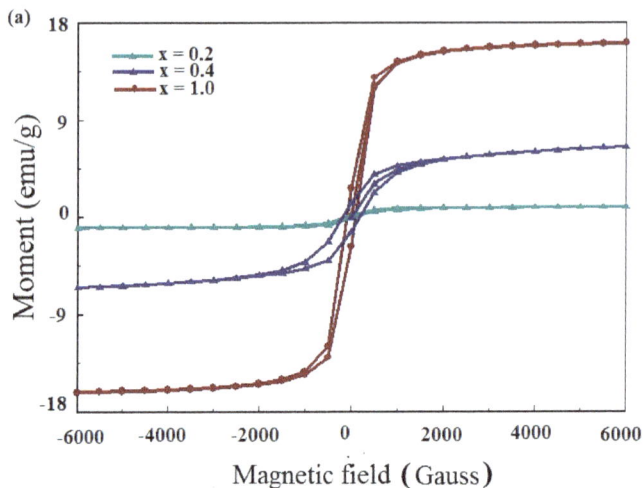

Figure 2.35 (a) *The magnetic characterization (M-H) of $(1-x)BaTiO_3 + xMgFe_2O_4$ for x = 0.2, 0.4, and 1.0*

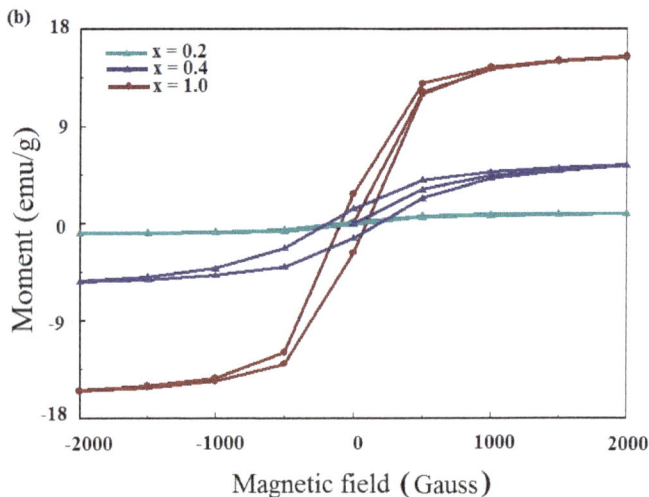

Figure 2.35 (b) *Expanded M-H curves of $(1-x)BaTiO_3 + xCoFe_2O_4$ for x = 0.2, 0.4, and 1.0*

Table 2.23 *Magnetic parameters of (1-x)BaTiO₃ + xMgFe₂O₄ composite from the study of magnetic hysteresis*

Composition x	Coercivity H_{ci} (G)	Magnetization M_s (emu/g)	Retentivity $M_r \times 10^{-3}$ (emu)
x = 0.2	93.073	1.500	7.164
x = 0.4	185.56	6.988	52.98
x = 1	92.764	16.451	136

2.8   MEM analysis: Charge density distribution

The electronic density of the prepared ceramic composite has been analyzed in the three-dimensional for the structure, two-dimensional and one-dimensional for the chemical bonding between the atoms forming the bonds. The bonding of the constituent atoms in the unit cell of the composite have been analyzed by maximum entropy method (MEM) [Collins, 1982] using the software PRIMA [Izumi, 2002] and VESTA [Momma, 2008].

2.8.1   (1-x)BaTiO₃ + xNiFe₂O₄

The electron densities of $BaTiO_3$ and $NiFe_2O_4$ are investigated separately. The three-dimensional electron density of $BaTiO_3$ in $(1-x)BaTiO_3$ + $xNiFe_2O_4$ ceramic composite is presented in Figures 2.36 (a) and (b). The two-dimensional electron density distribution between Ba-O for the compositions x = 0.2, 0.4, 0.6 and 0.8 is depicted in Figure 2.37 and the electron density for Ti-O bond is shown in Figure 2.38. The one-dimensional electron density profile of the Ba-O bond and Ti-O bond for all the compositions are shown in Figure 2.39 (a) and (b). The one-dimensional electron density values are presented in Table 2.24.

The three-dimensional electron density of $NiFe_2O_4$ in $(1-x)BaTiO_3$ + $xNiFe_2O_4$ ceramic composite is presented in Figures 2.40 (a) and (b). The two-dimensional electron density distribution between Ni-O and Fe-O for the compositions x = 0.2, 0.4, 0.6 and 0.8 are presented in Figure 2.41. The one-dimensional electron density profile of the Fe-O bond and Ni-O bond for all the compositions are shown in Figure 2.42 (a) and (b). The one-dimensional electron density values are presented in Table 2.25

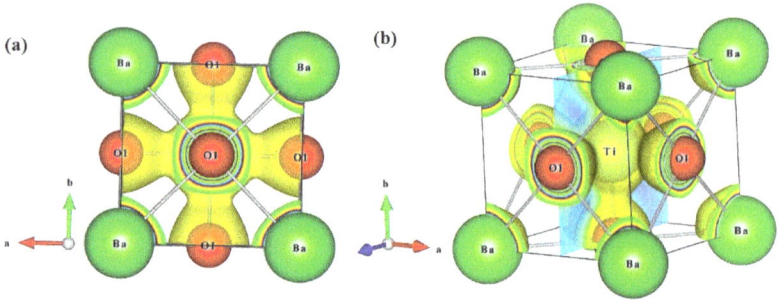

Figure 2.36 *Three-dimensional unit cell of BaTiO$_3$ of the composite (1-x) BaTiO$_3$ + x NiFe$_2$O$_4$ with isosurface level (a) 0.45 e/Å^3 (b) 0.8 e/Å^3 with plane (2 0 0)*

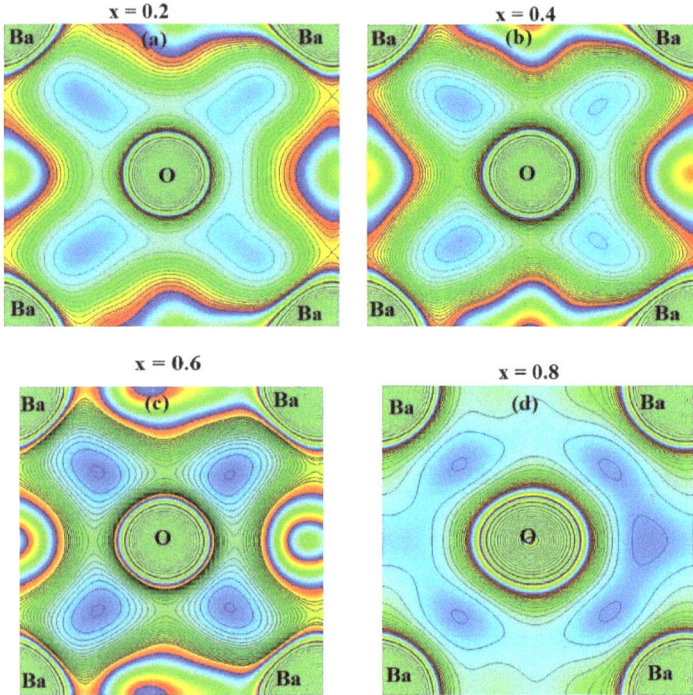

Figure 2.37 *The two-dimensional electron density distribution between Ba-O for the compositions x = 0.2, 0.4, 0.6 and 0.8.*

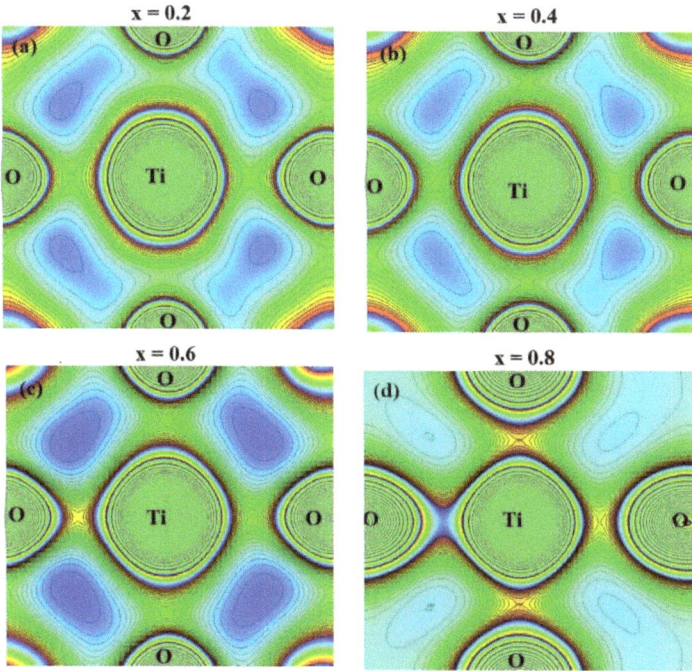

Figure 2.38 *The two-dimensional electron density distribution between Ti-O for the compositions x = 0.2, 0.4, 0.6 and 0.8.*

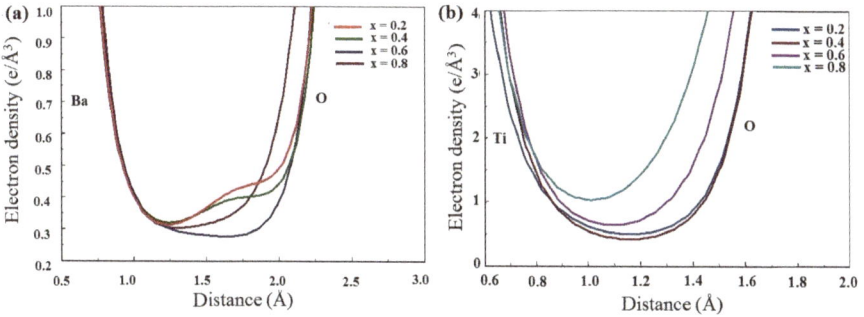

Figure 2.39 *One-dimensional electron density profile along (a) Ba–O bond, (b) Ti–O bond in BaTiO₃ of (1-x) BaTiO₃ + x NiFe₂O₄ composite for x = 0.2, 0.4, 0.6 and 0.8 compositions.*

Table 2.24 *Bond lengths and electron densities at bond critical point for Ti-O and Ba-O of the prepared composite for x = 0.2, 0.4, 0.6, 0.8 compositions from the one dimensional electron density profile for BaTiO₃ phase.*

x	Ti-O Bond Length (Å)	Electron density at bond critical point (e/Å³)	Ba-O Bond Length (Å)	Electron density at bond critical point(e/Å³)
0.2	2.0159	0.5153	2.8273	0.3139
0.4	2.0166	0.4149	2.8276	0.3209
0.6	2.0143	0.6369	2.8249	0.2771
0.8	2.0166	1.046	2.8259	0.3025

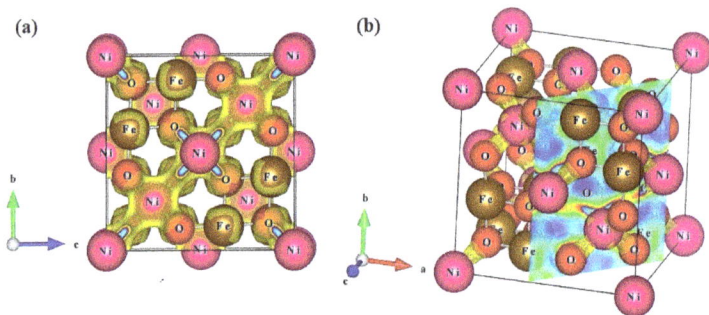

Figure 2.40 *Three-dimensional unit cell of NiFe₂O₄ of the composite (1-x) BaTiO₃ + x NiFe₂O₄ with isosurface level (a) 0.7 e/Å³ (b) 1.7 e/Å³ with plane (2 0 0)*

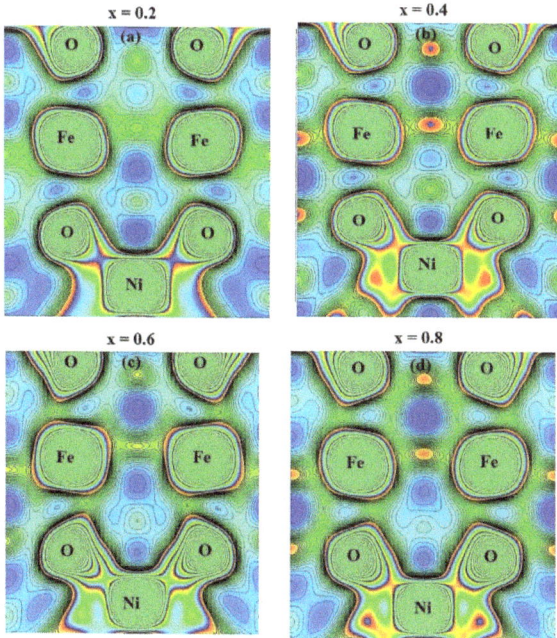

Figure 2.41 Two-dimensional charge density distribution between Ni–O and Fe–O of (1-x) BaTiO₃ + x NiFe₂O₄ on (1 0 1) plane contour lines varying from 0 – 0.8 e/Å³ with an interval of 0.04 e/Å³ (a) x = 0.2, (b) x = 0.4, (c) x = 0.6, and (d) x = 0.8

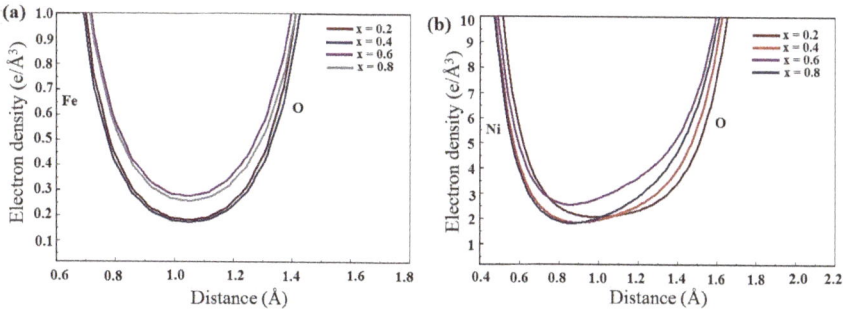

Figure 2.42 One-dimensional electron density profile along (a) Fe–O bond, (b) Ni–O bond in BaTiO₃ of (1-x) BaTiO₃ + x NiFe₂O₄ composite for x = 0.2, 0.4, 0.6 and 0.8 compositions.

Table 2.25 *Bond lengths and bond critical point electron densities for Ni-O and Fe-O of the prepared composite for x = 0.2, 0.4, 0.6, 0.8 compositions from the one dimensional electron density profile for NiFe$_2$O$_4$ phase*

x	Ni-O Bond Length (Å)	Electron density at bond critical point (e/Å^3)	Fe-O Bond Length (Å)	Electron density at bond critical point (e/Å^3)
0.2	1.9750	2.0890	1.9860	0.1831
0.4	1.9783	1.8836	1.9892	0.1623
0.6	1.9783	2.5918	1.9892	0.2740
0.8	1.9797	1.8227	1.9907	0.2514

2.8.2   (1-x)BaTiO$_3$ + xZnFe$_2$O$_4$

The electron densities of BaTiO$_3$ and ZnFe$_2$O$_4$ are investigated separately. The three-dimensional electron density of BaTiO$_3$ in (1-x)BaTiO$_3$ + xZnFe$_2$O$_4$ ceramic composite is presented in Figures 2.43 (a) and (b). The two-dimensional electron density distribution between Ba-O for the compositions x = 0.2, 0.4, 0.6 and 0.8 is depicted in Figure 2.44 and the electron density for Ti-O bond is shown in Figure 2.45. The one-dimensional electron density profile of the Ba-O bond and Ti-O bond for all the compositions are shown in Figure 2.46 (a) and (b). The one-dimensional electron density values are presented in Table 2.26.

The three-dimensional electron density of ZnFe$_2$O$_4$ in (1-x)BaTiO$_3$ + xZnFe$_2$O$_4$ ceramic composite is presented in Figures 2.47 (a) and (b). The two-dimensional electron density distribution between Zn-O and Fe-O for the compositions x = 0.2, 0.4, 0.6 and 0.8 are presented in Figure 2.48. The one-dimensional electron density profile of the Fe-O bond and Zn-O bond for all the compositions are shown in Figure 2.49 (a) and (b). The one-dimensional electron density values are presented in Table 2.27.

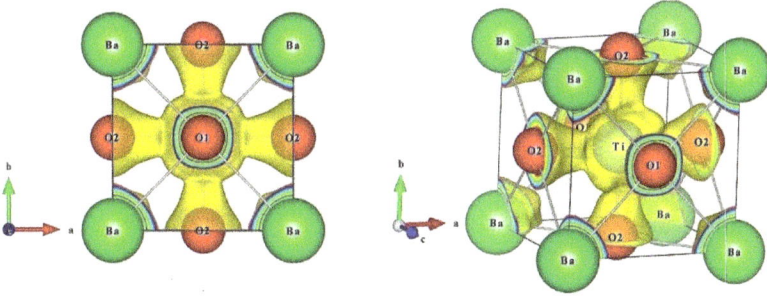

Figure 2.43 *Three-dimensional view of the structure of BaTiO₃ of the prepared composite in two different orientations with isosurface 0.9 e/Å³*

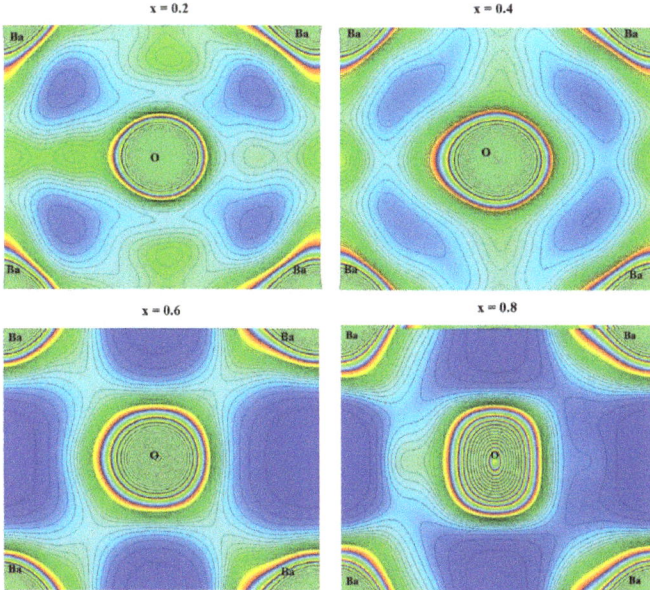

Figure 2.44 *Two-dimensional view of the electron density of Ba–O bond in BaTiO₃ of the prepared composite with contour interval 0.04 e/Å³ from 0 – 0.8e/Å³ for x = 0.2, 0.4, 0.6, and 0.8.*

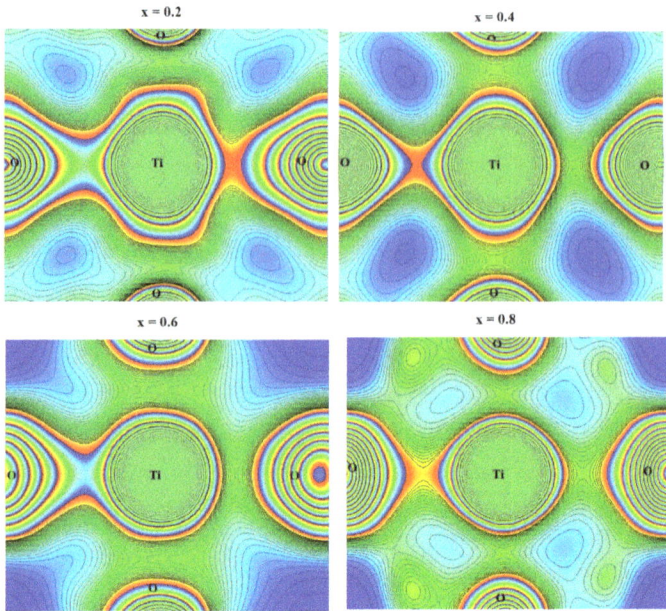

Figure 2.45 *Two-dimensional view of the electron density of Ti–O bond in BaTiO₃ of the prepared composite with compositions x = 0.2, 0.4, 0.6 and 0.8; contour interval 0.04 e/Å³ from 0 – 0.8 e/Å³*

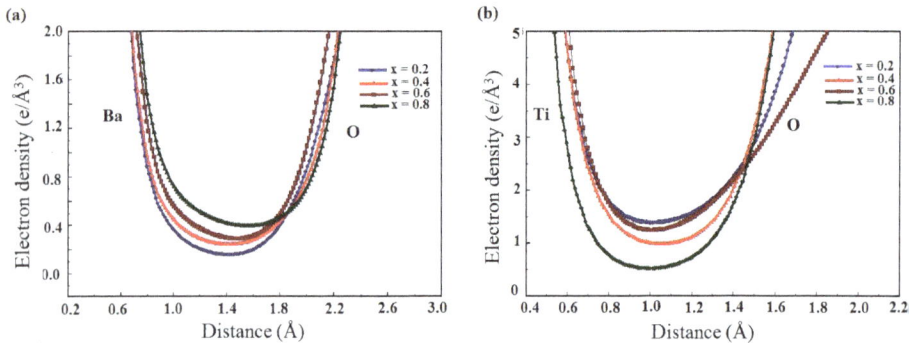

Figure 2.46 *One-dimensional electron density profile of Ba-O and Ti-O in BaTiO₃ of the prepared composite for x = 0.2, 0.4, 0.6 & 0.8.*

Table 2.26 *Bond lengths and electron densities at bond critical point for Ti-O and Ba-O of the prepared composite from the one-dimensional electron density profile for BaTiO₃ phase.*

x	Ti-O Bond Length (Å)	Electron density at bond critical point (e/Å³)	Ba-O Bond Length (Å)	Electron density at bond critical point(e/Å³)
0.2	2.0140	1.408	2.8324	0.145
0.4	2.0134	0.973	2.8298	0.234
0.6	2.0120	1.235	2.8287	0.285
0.8	2.0115	0.482	2.8257	0.385

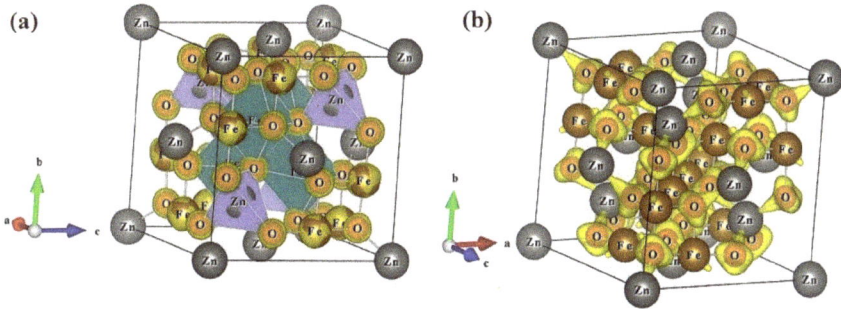

Figure 2.47 *Three-dimensional view of the structure of ZnFe₂O₄ of the prepared composite with isosurface 1.8 e/Å³ (a) Shows the polygons associated with Zn and Fe (b) in different orientation*

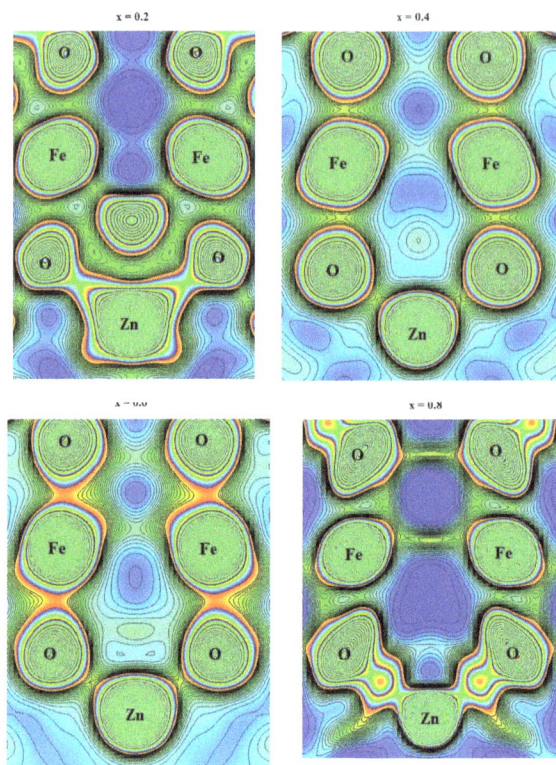

Figure 2.48 *Two-dimensional view of the electron density of Zn–O and Fe–O bonds in ZnFe$_2$O$_4$ of the prepared composite with contour interval 0.04 e/Å^3 from 0 – 0.8 e/Å^3.*

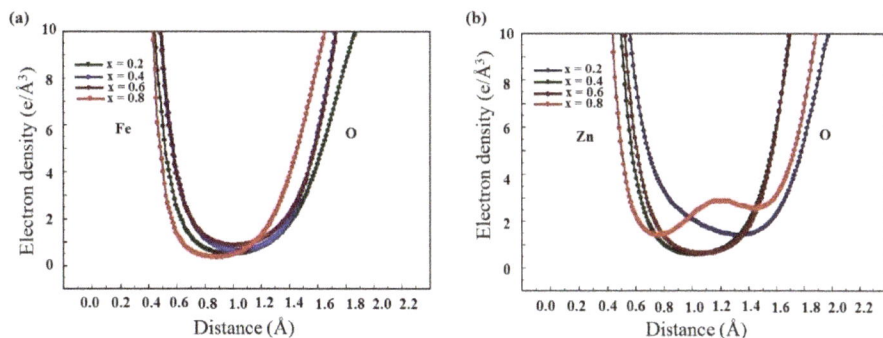

Figure 2.49 *One-dimensional electron density profile of Fe-O and Zn-O in ZnFe$_2$O$_4$ (x =*
0.2, 0.4, 0.6 & 0.8)

Table 2.27 *Bond lengths and bond critical point electron densities for Zn-O and Fe-O of*
the prepared composite from the one-dimensional electron density profile for ZnFe$_2$O$_4$
phase

x	Zn-O Bond Length (Å)	Electron density at bond critical point (e/Å^3)	Fe-O Bond Length (Å)	Electron density at bond critical point (e/Å^3)
0.2	2.0049	1.418	2.0160	0.523
0.4	2.0039	0.596	2.0150	0.673
0.6	2.0063	0.662	2.0173	0.864
0.8	2.0064	2.815	2.0175	0.353

2.8.3   (1-x)BaTiO$_3$ + xCoFe$_2$O$_4$

The electron densities of BaTiO$_3$ and CoFe$_2$O$_4$ are investigated separately. The three-dimensional electron density of BaTiO$_3$ in (1-x)BaTiO$_3$ + xCoFe$_2$O$_4$ ceramic composite is presented in Figures 2.50 (a) and (b). The two-dimensional electron density distribution between Ba-O for the compositions x = 0.2, 0.4, 0.6 and 0.8 is depicted in Figure 2.51 and the electron density for Ti-O bond is shown in Figure 2.52. The one-dimensional electron density profile of the Ba-O bond and Ti-O bond for all the compositions are shown in Figure 2.53 (a) and (b). The one-dimensional electron density values are presented in Table 2.28.

The three-dimensional electron density of CoFe$_2$O$_4$ in (1-x)BaTiO$_3$ + xCoFe$_2$O$_4$ ceramic composite is presented in Figures 2.54 (a) and (b). The two-dimensional electron density distribution between Co-O and Fe-O for the compositions x = 0.2, 0.4, 0.6 and 0.8 are presented in Figure 2.55. The one-dimensional electron density profile of the Fe-O bond and Co-O bond for all the compositions are shown in Figure 2.56 (a) and (b). The one-dimensional electron density values are presented in Table 2.29.

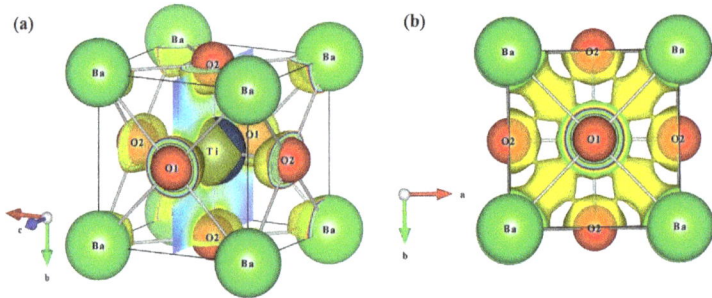

Figure 2.50 *Three-dimensional unit cell of BaTiO₃ of the composite (1-x) BaTiO₃ + x CoFe₂O₄ with isosurface level (a) 0.8 e/Å³ with plane (2 0 0) (b) 0.45 e/Å³.*

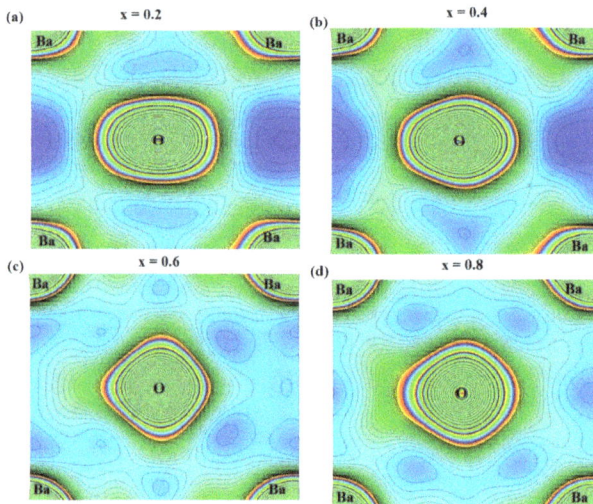

Figure 2.51 *Two-dimensional charge density distribution between Ba–O on (1 0 0) plane of BaTiO₃ of the composite (1-x) BaTiO₃ + x CoFe₂O₄ for x = 0.2, 0.4, 0.6, 0.8 with contour lines varying from 0 – 1 e/Å³ with an interval of 0.08 e/Å³.*

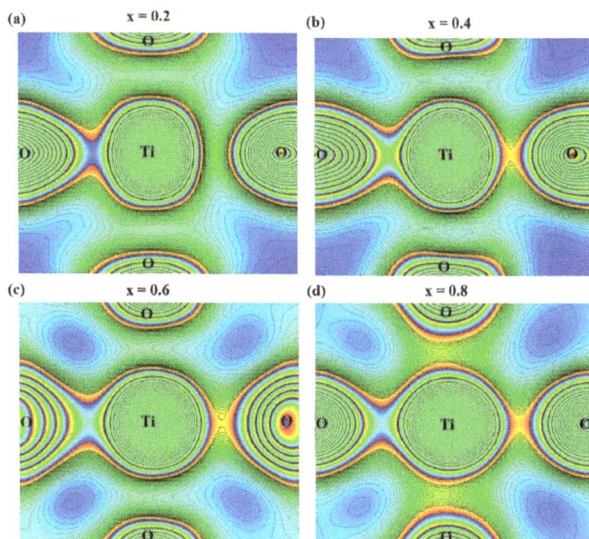

Figure 2.52 *Two-dimensional charge density distribution between Ti–O on (2 0 0) plane of BaTiO₃ of the composite (1-x) BaTiO₃ + x CoFe₂O₄ for x = 0.2, 0.4, 0.6, 0.8, with contour lines varying from 0 – 1 e/Å³ with an interval of 0.08 e/Å³.*

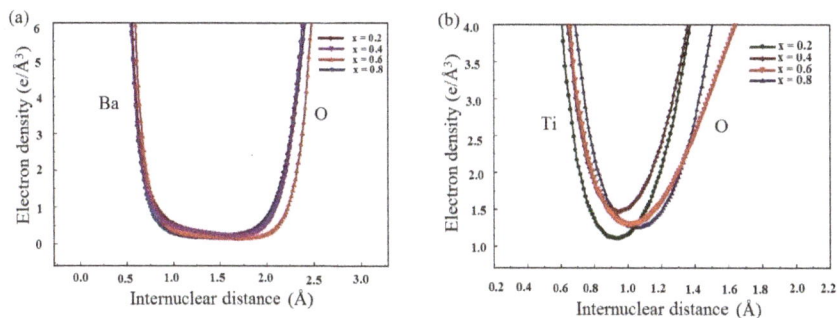

Figure 2.53 *One-dimensional electron density profiles along (a) Ba–O and (b) Ti–O bonds in BaTiO₃ of the prepared composite for x = 0.2, 0.4, 0.6 and 0.8 compositions.*

Table 2.28 *Bond lengths and electron densities at bond critical point for Ti-O and Ba-O of the prepared composite from the one-dimensional electron density profile for BaTiO₃ phase.*

x	Ti-O Bond Length (Å)	Electron density at bond critical point (e/Å^3)	Ba-O Bond Length (Å)	Electron density at bond critical point(e/Å^3)
0.2	2.0140	1.079	2.8382	0.253
0.4	2.0132	1.464	2.8375	0.233
0.6	2.0115	1.288	2.8364	0.148
0.8	2.0095	1.239	2.8340	0.177

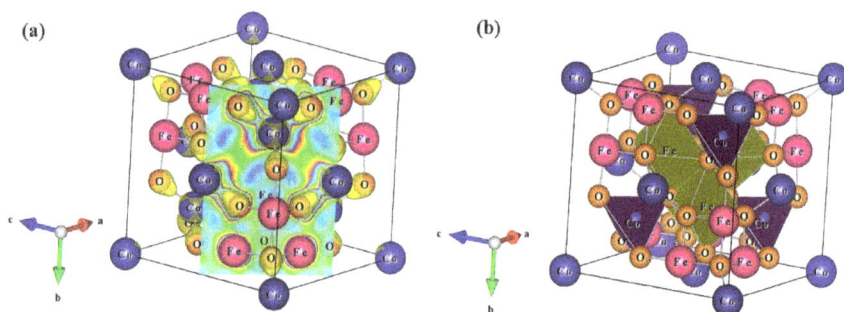

Figure 2.54 *Three-dimensional unit cell of CoFe₂O₄ of the composite (1-x) BaTiO₃ + x CoFe₂O₄ with (a) isosurface level of 1.8 e/Å^3 with plane (1 0 1) at 0.5d (b) the polygons associated with Co (tetrahedral), Fe (Octahedral).*

(a)

(b)

(c)

(d)

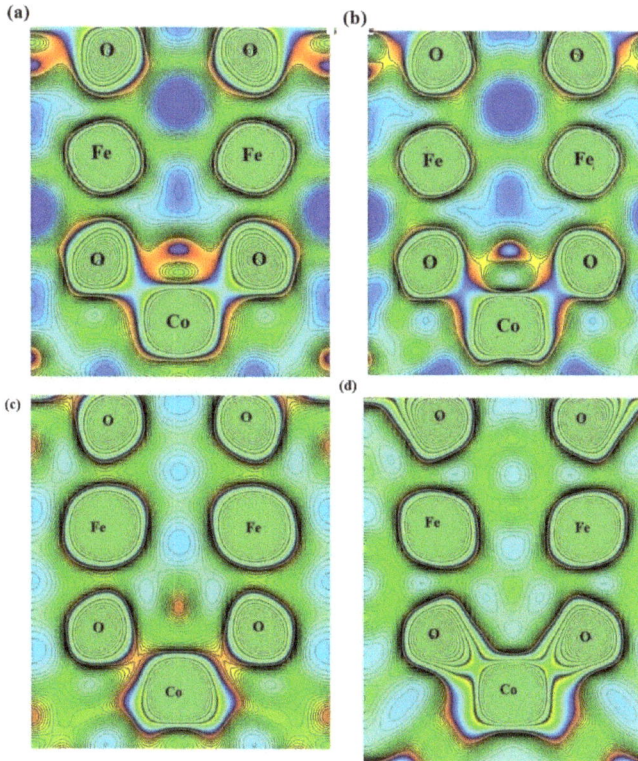

Figure 2.55 *Two-dimensional charge density distribution of Co–O and Fe–O on (1 0 1) plane of CoFe$_2$O$_4$, of (1-x) BaTiO$_3$ + x CoFe$_2$O$_4$ (a) x = 0.2 (b) x = 0.4 (c) x = 0.6 (d) x = 0.8 with contour lines varying from 0 to 1.8 e/Å^3 with an interval of 0.08 e/Å^3.*

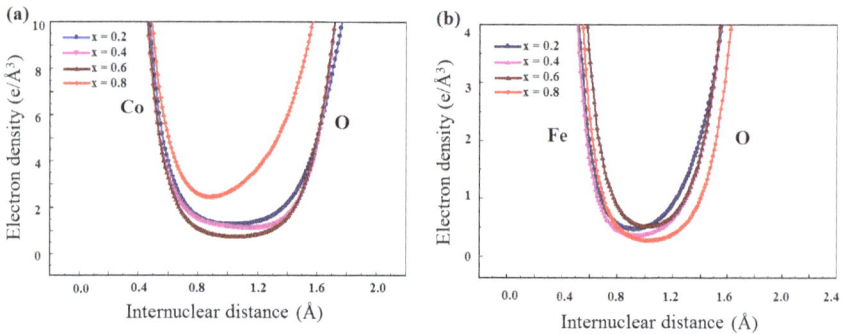

Figure 2.56 *One dimensional electron density profiles along (a) Co–O and (b) Fe–O bonds in CoFe$_2$O$_4$ of the prepared composite.*

Table 2.29 *Bond lengths and bond critical point electron densities for Co-O and Fe-O of the prepared composite from the one dimensional electron density profile for CoFe$_2$O$_4$ phase*

Composition x	Co-O Bond Length (Å)	Electron density at bond critical point (e/Å^3)	Fe-O Bond Length (Å)	Electron density at bond critical point (e/Å^3)
0.2	1.9891	1.284	2.0000	0.467
0.4	1.9905	1.096	2.0016	0.366
0.6	1.9904	0.718	2.0014	0.493
0.8	1.9902	2.391	2.0012	0.239

2.8.4 (1-x)BaTiO$_3$ + xMgFe$_2$O$_4$

The electron densities of BaTiO$_3$ and MgFe$_2$O$_4$ are investigated separately. The three-dimensional electron density of BaTiO$_3$ in (1-x)BaTiO$_3$ + xMgFe$_2$O$_4$ ceramic composite is presented in Figures 2.57 (a) and (b). The two-dimensional electron density distribution between Ba-O for the compositions x = 0.2, and 0.4 is depicted in Figure 2.58 and the electron density for Ti-O bond is shown in Figure 2.59. The one-dimensional electron density profile of the Ba-O bond and Ti-O bond for all the compositions are shown in Figure 2.60 (a) and (b). The one-dimensional electron density values are presented in Table 2.30

The three-dimensional electron density of $MgFe_2O_4$ in $(1-x)BaTiO_3 + xMgFe_2O_4$ ceramic composite is presented in Figures 2.61 (a) and (b). The two-dimensional electron density distribution between Mg-O and Fe-O for the compositions x = 0.2, 0.4, 0.6 and 0.8 are presented in Figure 2.62. The one-dimensional electron density profile of the Fe-O bond and Mg-O bond for all the compositions are shown in Figure 2.63 (a) and (b). The one-dimensional electron density values are presented in Table 2.31.

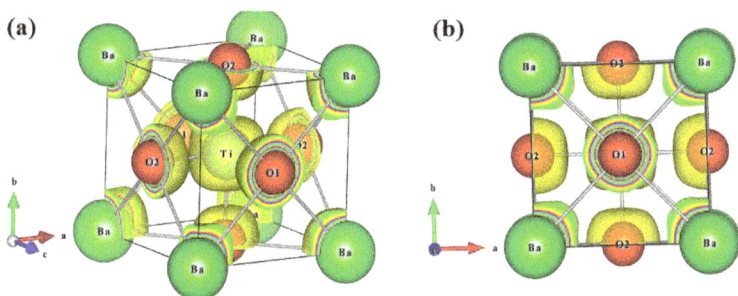

Figure 2.57 *Three-dimensional view of the structure of BaTiO3 of the prepared composite with an isosurface 0.5 e/Å³ in two different orientations (a) and (b).*

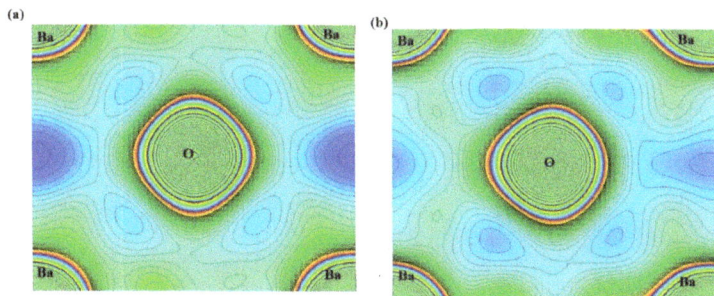

Figure 2.58 *Two-dimensional electron density distribution of Ba–O bond (a) x = 0.2, (b) x = 0.4 and Ti–O bond (c) x = 0.2, (d) x = 0.4 in BaTiO₃ of the prepared composite with contour interval 0.04 e/Å³ and contour lines vary from 0 – 0.8e/Å³*

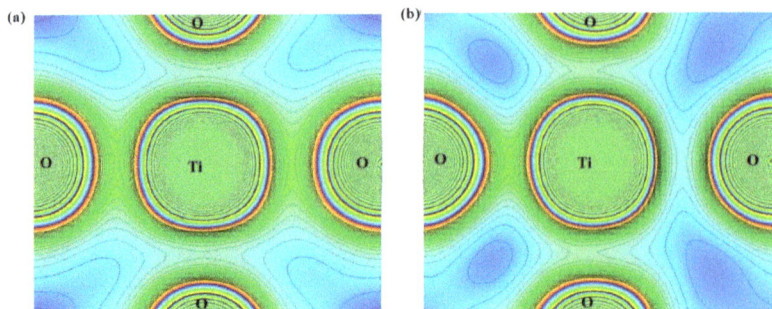

Figure 2.59 *Two-dimensional electron density distribution of Ti–O bond (a) x = 0.2, (b) x = 0.4 in BaTiO₃ of the prepared composite with contour interval 0.04 e/Å³ and contour lines vary from 0 – 0.8e/Å³*

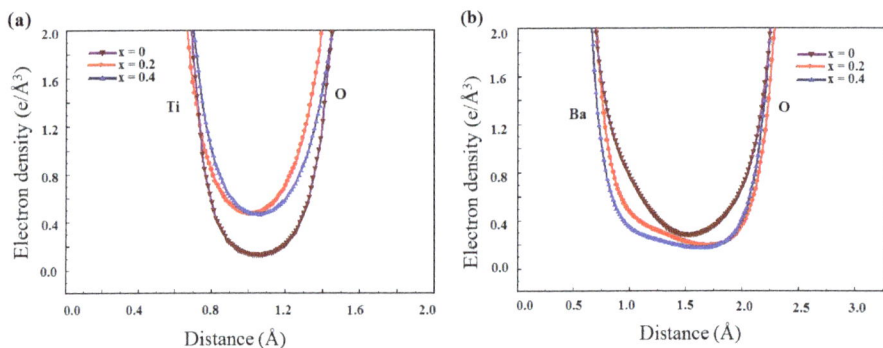

Figure 2.60 *One-dimensional electron density profile of (a) Ti–O and (b) Ba–O in BaTiO₃ for x = 0, 0.2, 0.4.*

Table 2.30 *Bond lengths and electron densities at bond critical point of Ti-O and Ba-O bonds of BaTiO$_3$ phase.*

Composition x	Ti-O Bond Length (Å)	Electron density at Bond critical point (e/Å^3)	Ba-O Bond Length (Å)	Electron density at bond critical point(e/Å^3)
0	2.021	0.113	2.830	0.287
0.2	2.011	0.470	2.832	0.202
0.4	2.014	0.471	2.838	0.154

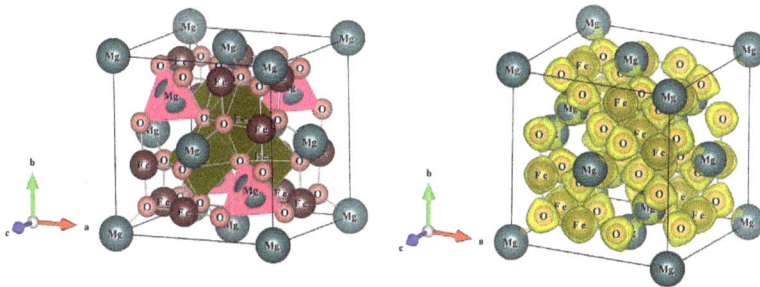

Figure 2.61 *Three-dimensional view of the structure of MgFe$_2$O$_4$ of the prepared composite (a) polygons associated with Mg and Fe (b) with an isosurface level 0.6 e/Å^3*

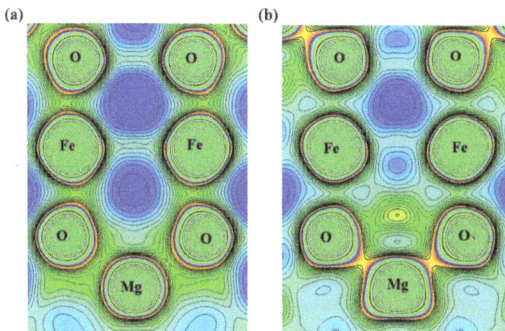

Figure 2.62 *Two-dimensional electron density distribution of Mg-O and Fe-O bonds in MgFe$_2$O$_4$ of the prepared composite with contour interval 0.04 e/Å^3 and contour lines vary from 0 – 0.8 e/Å^3.*

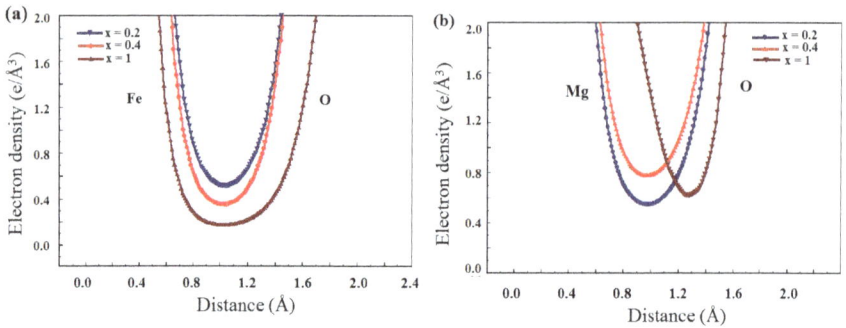

Figure 2.63 *One-dimensional electron density profile of Fe−O and Mg−O in MgFe₂O₄ (x = 0.2, 0.4, 1)*

Table 2.31 *Bond lengths and electron densities at bond critical point of Mg - O and Fe - O of MgFe₂O₄ phase*

Compositiion x	Mg-O Bond Length (Å)	Electron density at bond critical point (e/Å³)	Fe-O Bond Length (Å)	Electron density at bond critical point (e/Å³)
0.2	2.006	0.488	2.017	0.528
0.4	1.989	0.706	2.000	0.337
1	1.981	0.599	1.992	0.165

References

• Collins D. M. Nature 298 49 (1982) https://doi.org/10.1038/298049a0

• Hollinsworth B. S, Mazumdar D, Sims H, et al., Chemical tuning of optical bandgap in spinel ferrites: CoFe2O4 Vs NiFe2O4 Appl. Phys. Lett. 103 082406 (2013) https://doi.org/10.1063/1.4818315

• Izumi F, Dilanien R.A, Recent Research Developments in Physics Part II, Vol. 3, Transworld Research Network. Trivandrum. 699 (2002).

• Momma K, Izumi F, VESTA: a three-dimensional visualization system for electronic and structural analysis. J. Appl. Crystallogr 41 653 (2008) https://doi.org/10.1107/S0021889808012016

- Petricek V, Dusek M, Palatinus L, Kristallogr Z, Crystallographic Computing system JANA 2006: General features 229 345 (2014) https://doi.org/10.1515/zkri-2014-1737

- Rietveld H. M, J. Appl. Crystallogr. 2 65 (1969) https://doi.org/10.1107/S0021889869006558

- Saravanan R, GRAIN software, Personal communication (2008)

- Wood D. L, Tauc J, Phys Rev B. 5 3144 (1972). https://doi.org/10.1103/PhysRevB.5.3144

Chapter 3

Analysis of the Results

Abstract

This chapter gives a detailed analysis of all the characterizations observed for the prepared composites. The details of sample preparation for different composites are discussed. The structural analysis of the prepared composites using powder XRD data is analyzed for all the composites. The morphological nature of the composites is analyzed from the SEM images. Also gives the details of particle and grain sizes of the prepared composites. The elementary compositions of the composites have been discussed from the EDS data of all the composites. The optical characterization from the UV-Vis spectral data is analyzed from the bandgap. The electrical characterization; dielectric, dielectric loss, and P-E hysteresis are discussed from the corresponding data. The magnetic characterization of the composites from M-H hysteresis data are analyzed. The electron density studies depicting the electronic structure, nature of bonding, electron densities correlation with the observed properties are explained. The analysis of the observed results, correlation with the electron density studies is discussed in detail.

Keywords

Composite Material, Two Phase Analysis, Structural, Morphological, P-E hysteresis, M-H Hysteresis

3.1 Introduction

In the present work, four series of magneto-electric ceramic composite has been synthesized by solid state reaction method. The prepared ceramic composites are characterized by powder X-ray diffraction (PXRD), scanning electron microscope (SEM), electron dispersive X-ray spectroscopy (EDS), Optical characterization: UV-Visible spectroscopy, Electrical characterization: dielectric/ capacitance versus frequency, P-E hysteresis studies, Magnetic characterization: M-H hysteresis studies. By applying different analytical techniques; Rietveld refinement technique [Rietveld, 1969], Maximum Entropy Method (MEM) [Collins, 1982], the electronic structure, charge density distribution and chemical bonding of the prepared ceramic composites have been investigated. The results from the experimental characterization methods and analytical techniques are presented in chapter 2.

The detailed analysis of the four series of magneto-electric ceramic composites is discussed in chapter 3. The sample preparation details of all the four series are discussed in **section**

Materials Research Forum LLC
https://doi.org/10.21741/9781644902196

3.2. The structural characterization of the prepared composites is carried out using powder X-ray diffraction. All the experimental data of PXRD are refined through Rietveld refinement [Rietveld, 1969] using the software JANA 2006 [Petricek et al., 2014]. The refined structural data of all the four series are compared in **section 3.3.** The morphological and the microstructure of the prepared composites are analyzed from the scanning electron microscopy (SEM) images. The particle sizes and the grain sizes of the composites have been discussed in **section 3.4.** The GRAIN software [Saravanan, Personal communication] is used to calculate the grain sizes of $BaTiO_3$ and the ferrites forming the ferrite-ferroelectric ceramic composites. The elementary compositions of the prepared composites have been analyzed quantitatively and qualitatively from the results of EDS data and spectrum. The atomic and weight percentage of all the composite series are discussed in **section 3.5.**

The optical characterization from the UV-Visible spectral data is discussed for all the samples of the composites in **section 3.6.** The optical band gap values have been evaluated using Tauc plot methodology [Wood and Tauc, 1972]. The variation of optical bandgap value with the composition in a series is also discussed clearly.

The electrical characterization is discussed from the results of the capacitance versus frequency graph, the dielectric loss and the P-E characterization of all the series of the prepared composites in **section 3.7.**

The magnetic characterization is discussed from the M-H data extracted from the vibrating signal magnetometer (VSM) in **section 3.8.**

The electronic structure, chemical bonding between the atoms and the charge density distribution in the unit cell of the prepared samples have been analyzed by maximum entropy method (MEM) [Collins, 1982] using the softwares PRIMA [Izumi, 2002] and VESTA [Momma, 2008]. The 3D electron density iso-surfaces, 2D electron density contour maps and 1D electron density line profiles for all the prepared composites for all compositions $x = 0.2, 0.4, 0.6, 0.8$ are elaborately explained in **section 3.9.**

3.2   Sample preparation

In this work four series of magneto-electric ceramic composites were prepared by solid state sintering method. $BaTiO_3$ is the ferroelectric material which is considered for all the four series. Ferrites which were used to form the composites are MFe_2O_4 where M = Ni, Zn, Co and Mg. Hence the four series are (1-x) $BaTiO_3$ + $xNiFe_2O_4$, (1-x) $BaTiO_3$ + $xZnFe_2O_4$, (1-x) $BaTiO_3$ + $xCoFe_2O_4$, (1-x) $BaTiO_3$ + $xMgFe_2O_4$.

Preparation of BaTiO$_3$

The ferroelectric phase (BaTiO$_3$) was synthesized by taking suitable quantities of BaCO$_3$ (Alfa Aesar99.99%) and TiO$_2$ (Alfa Aesar 99.99%) in stoichiometric proportion and mixing them thoroughly using an agate mortar for 4 h. The mixture was then pelletized and sintered for 2 h at 1350°C in the air. Preparation of ferrites and BaTiO$_3$ are tabulated for comparison in table 3.1.

Table 3.1 *The synthesis of ferrites and BaTiO$_3$ with sintering temperature & duration*

SAMPLES	GRINDING DURATION (HOURS)	SINTERING TEMPERATURE (°C)	SINTERING DURATION (HOURS)
BATIO$_3$	4	1350	2
COFE$_2$O$_4$	12	1100	12
MGFE$_2$O$_4$	12	1200	5

The composites of the ferrite-ferroelectric were synthesized by taking the samples according to the formula (1-x) BaTiO$_3$ + x MFe$_2$O$_4$ (M; Ni, Zn, Co, Mg). The preparation details are given in table 3.2.

Table 3.2 *Comparison of the synthesis of composites with sintering temperature & duration*

Samples	Grinding duration (hours)	Sintering Temperature (°C)	Sintering Duration (hours)
(1-x)BaTiO$_3$ + xNiFe$_2$O$_4$	12	1250	5
(1-x)BaTiO$_3$ + xZnFe$_2$O$_4$	12	1200	5
(1-x)BaTiO$_3$ + xCoFe$_2$O$_4$	12	1150	12
(1-x)BaTiO$_3$ + xMgFe$_2$O$_4$	12	1100	15

The grinding hours are the same 12 hours for all the series, the sintering temperature and the duration are different since it depends on the reactivity nature of the ferrite. In a series, the value of x is 0.2, 0.4, 0.6 and 0.8.

3.3 Structural analysis of the magneto-electric ceramic composites

The structural characterization of the prepared magneto-electric ceramic composites has been analyzed from the powder XRD data set. The XRD data for all the samples were collected for $2\theta = 10°$ to $120°$ and refined by Rietveld method [Rietveld, 1969] using JANA 2006 software [Patricek et al., 2014]. The position coordinates of the atoms of the $BaTiO_3$ and ferrites are found from the standard Wycoff position table [Wycoff, 1963]. The detailed discussion of the structure analysis is given in this section.

3.3.1   (1-x)BaTiO₃ + xNiFe₂O₄ (x = 0.2, 0.4, 0.6 and 0.8)

The prepared composite subjected to Rietveld refinement gives the structural parameters. The raw XRD profiles of the composite for all compositions (x = 0.2, 0.4, 0.6, 0.8) is shown in **figure 2.1**, the 2θ of the corresponding Bragg peaks match with Joint Committee on Powder Diffraction Standards (JCPDS). The peaks corresponding to both the phases are present at different angles (2θ) and some of the peaks correspond to both the phases The **JCPDS No. for BaTiO₃ phase is 05-0626** and **JCPDS No. 10-0325** is for the $NiFe_2O_4$ phase. $BaTiO_3$ is identified to have a tetrahedral perovskite structure with **space group P4mm (Space group No. 99)** and $NiFe_2O_4$ is cubic with **space group Fd-3m (Space group No. 227)**. As the ferrite content increases from x = 0.2 to x = 0.8, the intensity of the ferrite peaks increase and the intensity of the ferroelectric peak decreases.

The refined profiles from the Rietveld refinement [Rietveld, 1969] using JANA 2006 [Patricek et al., 2014] are shown in **figure 2.2 (a), (b), (c), (d) for x = 0.2, 0.4, 0.6, 0.8.** The profiles are refined for two phases $BaTiO_3$ and $NiFe_2O_4$. There are vertical lines below the profile are the Braggs peaks and there are two rows of vertical lines corresponding to two phases. During refinement the observed peaks are matched with the calculated peaks and the difference between the observed and the calculated are shown as the error peaks below the vertical lines. The line of the error peaks shows that the profiles are well refined. The refined parameters are tabulated in **table 2.1**. From **table 2.1** it can be observed that the lattice parameter of the ferroelectric $BaTiO_3$ is maximum for x = 0.4 and minimum for x = 0.6. The c/a ratio for x = 0.4 is minimum (1.0076) and maximum for x = 0.8 (1.0091). For pure $BaTiO_3$ the c/a ratio is 1.011. The c/a ratio of the ferroelectric plays a role in its electrical property.

The lattice parameter of the $NiFe_2O_4$ is a maximum for x = 0.6 which increases from x = 0.2 (a minimum). When the lattice parameter changes in magnitude, the volume of the crystal changes, either increases or decrease which shows the influence of ferrite on ferroelectric and vice versa.

3.3.2   (1-x)BaTiO₃ + xZnFe₂O₄ (x = 0.2, 0.4, 0.6 and 0.8)

The prepared composite subjected to Rietveld refinement gives the structural parameters. The raw XRD profiles of the composite for all compositions (x = 0.2, 0.4, 0.6, 0.8) is shown in **figure 2.3**, the 2θ of the corresponding Bragg peaks match with Joint Committee on

Powder Diffraction Standards (JCPDS). The peaks corresponding to both the phases are present at different angles (2θ) and some of the peaks correspond to both the phases The **JCPDS No.** for **BaTiO₃ phase is 05-0626** and **JCPDS No. 22-1012** is for the $ZnFe_2O_4$ phase. **BaTiO₃** is identified to have a tetrahedral perovskite structure with **space group P4mm (Space group No. 99)** and **ZnFe₂O₄** is cubic with **space group Fd-3m (Space group No. 227).** As the ferrite content increases from x = 0.2 to x = 0.8, the intensity of the ferrite peaks increase and the intensity of the ferroelectric peak decreases.

The refined profiles from the Rietveld refinement [Rietveld, 1969] using JANA 2006 [Patricek et al., 2014] are shown in **figures 2.4 (a), (b), (c), (d) for x = 0.2, 0.4, 0.6, 0.8.** The profiles are refined for two phases BaTiO₃ and ZnFe₂O₄. There are vertical lines below the profile are the Braggs peaks and there are two rows of vertical lines corresponding to two phases. During refinement the observed peaks are matched with the calculated peaks and the difference between the observed and the calculated are shown as the error peaks below the vertical lines. The line of the error peaks shows that the profiles are well refined. The refined parameters are tabulated in **table 2.2.** From table 2.2 it can be observed that the lattice parameter of the ferroelectric BaTiO₃ is maximum for x = 0.8 and minimum for x = 0.6. The c/a ratio for x = 0.2 is minimum (1.0055) and maximum for x = 0.4 (1.0062). For pure BaTiO₃ the c/a ratio is 1.011. The c/a ratio of the ferroelectric plays a role in its electrical property.

The lattice parameter of the ZnFe₂O₄ is a maximum for x = 0.2 and a minimum for x = 0.8. When the lattice parameter changes in magnitude, the volume of the crystal changes, either increases or decrease which shows the influence of ferrite on ferroelectric and vice versa.

3.3.3   (1-x)BaTiO₃ + xCoFe₂O₄ (x = 0.2, 0.4, 0.6 and 0.8)

The prepared composite subjected to Rietveld refinement gives the structural parameters. The raw XRD profiles of the composite for all compositions (x = 0.2, 0.4, 0.6, 0.8) is shown in **figure 2.5**, the 2θ of the corresponding Bragg peaks match with Joint Committee on Powder Diffraction Standards (JCPDS). The peaks corresponding to both the phases are present at different angles (2θ) and some of the peaks correspond to both the phases The **JCPDS No.** for **BaTiO₃ phase is 05-0626** and **JCPDS No. 22-1086** is for the $CoFe_2O_4$ phase. **BaTiO₃** is identified to have a tetrahedral perovskite structure with **space group P4mm (Space group No. 99)** and **CoFe₂O₄** is cubic with **space group Fd-3m (Space group No. 227).** As the ferrite content increases from x = 0.2 to x = 0.8, the intensity of the ferrite peaks increases and the intensity of the ferroelectric peak decreases. The refined profiles from the Rietveld refinement [Rietveld, 1969] using JANA 2006 [Patricek et al., 2014] are shown in **figures 2.6 (a), (b), (c), (d) for x = 0.2, 0.4, 0.6, 0.8.** The profiles are refined for two phases BaTiO₃ and CoFe₂O₄. There are vertical lines below the profile are the Braggs peaks and there are two rows of vertical lines corresponding to two phases. During refinement the observed peaks are matched with the calculated peaks and the difference between the observed and the calculated are shown as the error peaks below the vertical lines. The line of the error peaks shows that the profiles are well refined. The refined parameters are tabulated in **table 2.3**. From **table 2.3** it can be observed that the lattice parameter of the structure of the ferroelectric $BaTiO_3$ is maximum for x = 0.2 and minimum for x = 0.8. The c/a ratio for x = 0.8 is minimum (1.0055) and maximum for x = 0.2 (1.0072). For pure $BaTiO_3$ the c/a ratio is 1.011. The c/a ratio of the ferroelectric plays a role in its electrical property.

The lattice parameter structure of the $CoFe_2O_4$ is a maximum for x = 0.2 and a minimum for x = 0.8. When the lattice parameter changes in magnitude, the volume of the crystal changes, either increases or decrease which shows the influence of ferrite on ferroelectric and vice versa.

3.3.4   (1-x)BaTiO₃ + xMgFe₂O₄ (x = 0, 0.2, 0.4, 1)

The prepared composite subjected to Rietveld refinement gives the structural parameters. The raw XRD profiles of the composite for all compositions (x = 0.2, 0.4, 0.6, 0.8) is shown in **figure 2.6**, the 2θ of the corresponding Bragg peaks match with Joint Committee on Powder Diffraction Standards (JCPDS). The peaks corresponding to both the phases are present at different angles (2θ) and some of the peaks correspond to both the phases The **JCPDS No. for BaTiO₃ phase is 05-0626** and **JCPDS No. 36-0398** is for the $CoFe_2O_4$ phase. **BaTiO₃** is identified to have a tetrahedral perovskite structure with **space group P4mm (Space group No. 99)** and **CoFe₂O₄** is cubic with **space group Fd-3m (Space group No. 227)**. As the ferrite content increases from x = 0.2 to x = 0.8, the intensity of the ferrite peaks increases and the intensity of the ferroelectric peak decreases.

The refined profiles from the Rietveld refinement [Rietveld, 1969] using JANA 2006 [Patricek et al., 2014] are shown in **figures 2.8 (a), (b), (c), (d) for x = 0, 0.2, 0.4, 1.** The profiles are refined for two phases $BaTiO_3$ and $CoFe_2O_4$. There are vertical lines below the profile are the Braggs peaks and there are two rows of vertical lines corresponding to two phases. During refinement the observed peaks are matched with the calculated peaks and the difference between the observed and the calculated are shown as the error peaks below the vertical lines. The line of the error peaks shows that the profiles are well refined. The refined parameters are tabulated in **table 2.4**. From **table 2.4**, the lattice parameters of the structure of the ferroelectric $BaTiO_3$ are maximum for x = 0 and minimum for x = 0.2. Hence for pure $BaTiO_3$ the structure is maximum and it decreases with the increase in the ferrite and hence minimum for x = 0.4. The c/a ratio for x = 0 is minimum (1.0098) and maximum for x = 0.4 (1.0072). For pure $BaTiO_3$ c/a ratio is 1.011. The c/a ratio of the ferroelectric plays a role in its electrical property.

The lattice parameter of the structure of the $MgFe_2O_4$ is a maximum for x = 1 and a minimum for x = 0.2. Hence the structure of pure $MgFe_2O_4$ is a maximum. When the lattice parameter changes in magnitude, the volume of the crystal changes, either increases or decreases which shows the influence of ferrite on ferroelectric and vice versa.

Table 3.3 *Comparison of the refined structural parameters for $BaTiO_3$ and the ferrites*

Composites	Ferroelectric ($BaTiO_3$): Tetrahedral-P4mm				Ferrite:Cubic Fd-3m	
	a(Å)	c(Å)	Volume ($Å^3$)	c/a	a(Å)	Volume ($Å^3$)
BTNF x = 0.2	3.9985(0)	4.0318(0)	64.4609	1.0083	8.3233(0)	576.6202
x = 0.4	**4.0042(0)**	**4.0346(11)**	**64.626**	1.0076	8.3383(5)	579.7415
x = 0.6	3.9949(0)	4.0285(0)	64.2932	1.0084	**8.3449(8)**	**581.1196**
x = 0.8	3.9964(0)	4.0331(0)	64.4121	**1.0091**	8.3430(0)	580.7238
BTZF x = 0.2	4.0058(2)	4.0281(2)	64.636(5)	1.0055	**8.4492(0)**	**603.177(23)**
x = 0.4	4.0019(0)	4.0268(0)	64.488(0)	**1.0062**	8.4442(0)	602.100(0)
x = 0.6	4.0004(0)	4.0240(0)	64.398(0)	1.0059	8.4482(0)	602.976(0)
x = 0.8	**4.0070(0)**	**4.0314(0)**	**64.728(0)**	1.006	8.4424(0)	601.728(0)
BTCF x = 0.2	3.9995 (0)	4.0281 (1)	64.43 (0)	1.0072	**8.3894 (0)**	**590.47 (5)**
x = 0.4	3.9993 (5)	4.0264 (5)	64.40 (2)	1.0068	8.3868 (10)	589.93 (8)
x = 0.6	4.0003 (5)	4.0237 (6)	64.40 (2)	1.0058	8.3875 (9)	590.07 (8)
x = 0.8	3.9968 (0)	4.0189 (0)	64.20 (0)	1.0055	8.3778 (0)	588.03 (0)
BTMF x = 0.0	**4.0022 (2)**	**4.0413 (2)**	**64.73 (1)**	**1.0098**	-------	------
x = 0.2	3.9918 (0)	4.0255 (0)	64.14 (0)	1.0084	8.3729 (0)	586.99 (0)
x = 0.4	3.9989 (1)	4.0278 (2)	64.41 (1)	1.0072	8.3825 (0)	588.99 (2)
x = 1.0	------	------	------	----	8.4313 (9)	601.02 (04)

BTNF: $(1-x)BaTiO_3 + xNiFe_2O_4$; BTZF: $(1-x)BaTiO_3 + xZnFe_2O_4$; BTCF: $(1-x)BaTiO_3 + xCoFe_2O_4$; BTMF: $(1-x)BaTiO_3 + xMgFe_2O_4$

3.4   Morphological analysis of the composites

The surface morphology of all the prepared composites are analyzed from the SEM images of the composites. The SEM images are observed with different magnifications from x5000 to x25000. From the SEM images the average particle size is calculated.

3.4.1 $(1-x)BaTiO_3 + xNiFe_2O_4$

The SEM images of the $(1-x)BaTiO_3 + xNiFe_2O_4$ composite for the compositions x = 0.2, 0.4, 0.6, and 0.8 are shown in **figure 2.9**. The average particle size and the individual particle size of the prepared composite are given in **table 2.5**. The average particle size of the composite for compositions x = 0.2, 0.4, 0.6 and 0.8 varies from 1.74 to 3.40 μm. The individual particle size of $BaTiO_3$ and $NiFe_2O_4$ are found using a software GRAIN developed by Dr. R. Saravanan [Saravanan, 2008]. The individual particle size of $BaTiO_3$ varies from 23.52 to 32.29 nm and $NiFe_2O_4$ in the range of 16.24 to 35.27 nm.

3.4.2 $(1-x)BaTiO_3 + xZnFe_2O_4$

The SEM images of the $(1-x)BaTiO_3 + xNiFe_2O_4$ composite for the compositions x = 0.2, 0.4, 0.6, and 0.8 are shown in **figure 2.10**. The average particle size and the individual particle size of the prepared composite are given in **table 2.6**. The average particle size of the composite for compositions x = 0.2, 0.4, 0.6 and 0.8 varies from 1.021 to 1.575 μm. The individual particle size of $BaTiO_3$ and $ZnFe_2O_4$ are found using a software GRAIN developed by Dr. R. Saravanan [Saravanan, 2008]. The individual particle size of $BaTiO_3$ varies from 22.33 to 24.43 nm and $ZnFe_2O_4$ are in the range of 22.29 to 28.25 nm.

3.4.3 $(1-x)BaTiO_3 + xCoFe_2O_4$

The SEM images of the $(1-x)BaTiO_3 + xCoFe_2O_4$ composite for the compositions x = 0.2, 0.4, 0.6, and 0.8 are shown in **figure 2.11**. The average particle size and the individual particle size of the prepared composite are given in **table 2.7**. The average particle size of the composite for compositions x = 0.2, 0.4, 0.6 and 0.8 varies from 1.43 to 1.82 μm. The individual particle size of $BaTiO_3$ and $CoFe_2O_4$ are found using the software GRAIN developed by Dr. R. Saravanan [Saravanan, 2008]. The individual particle size of $BaTiO_3$ varies from 29.38 to 34.98 nm and $CoFe_2O_4$ are in the range of 32.29 to 39.42 nm.

3.4.4 $(1-x)BaTiO_3 + xMgFe_2O_4$

The SEM images of the $(1-x)BaTiO_3 + xMgFe_2O_4$ composite for the compositions x = 0.0 0.2, 0.4, 1.0 are shown in **figure 2.12**. The average particle size and the individual particle size of the prepared composite are given in **table 2.8**. The average particle size of the composite for compositions x = 0.0, 0.2, 0.4, and 1.0 varies from 0.966 to 1.384 μm. The individual particle size of $BaTiO_3$ and $MgFe_2O_4$ are found using a software GRAIN developed by Dr. R. Saravanan [Saravanan, 2008]. The individual particle size of $BaTiO_3$ varies from 17.46 to 26.01 nm and $MgFe_2O_4$ are in the range of 28.93 to 36.76 nm.

Table 3.4 Comparison of the average and the individual particle (grain) size

Composites	Avg. particle size (µm)	Individual particle size BaTiO3 (nm)	Individual particle size ferrite (nm)
BTNF	1.74 - 3.40	23.52 - 32.29	16.24 - 35.27
BTZF	1.02 - 1.57	22.33 - 24.43	22.29 - 28.25
BTCF	1.43 -1.82	29.38 - 34.98	32.29 - 39.42
BTMF	0.966 - 1.384	17.46 - 26.01	28.93 - 36.76

BTNF: $(1-x)BaTiO_3 + xNiFe_2O_4$; BTZF: $(1-x)BaTiO_3 + xZnFe_2O_4$; BTCF: $(1-x)BaTiO_3 + xCoFe_2O_4$; BTMF: $(1-x)BaTiO_3 + xMgFe_2O_4$

3.5   Elemental composition analysis

The elemental compositions of all the prepared composites of all compositions have been investigated from the energy dispersive X-ray spectroscopy (EDS). The EDS spectrum qualitatively verifies the chemical compositions of the prepared composites. The atomic and mass percentages of the doped titanates have also been verified by EDS quantitative analysis.

3.5.1 $(1-x)BaTiO_3 + xNiFe_2O_4$

Figure 2.13 shows the energy dispersive X-ray spectroscopy (EDS) spectra of the composite for compositions for x = 0.2, 0.4, 0.6 & 0.8. The elements present in the composite are only present which confirms that the composite is well prepared without impurities. The elemental compositions in terms of mass percentages and atomic percentages are listed in **table 2.9**. The atomic percentages and mass percentages of the prepared samples confirm the stoichiometry of the prepared samples.

3.5.2 $(1-x)BaTiO_3 + xZnFe_2O_4$

Figure 2.14 shows the energy dispersive X-ray spectroscopy (EDS) spectra of the composite **$(1-x)BaTiO_3 + xZnFe_2O_4$** for compositions for x = 0.2, 0.4, 0.6 & 0.8. The elements present in the composite are only present which confirms that the composite is well prepared without impurities. The elemental compositions in terms of mass percentages and atomic percentages are listed in **table 2.10**. The atomic percentages and mass percentages of the prepared samples confirm the stoichiometry of the prepared samples.

3.5.3   (1-x)BaTiO₃ + xCoFe₂O₄

Figure 2.15 shows the energy dispersive X-ray spectroscopy (EDS) spectra of the composite **(1-x)BaTiO₃ + xCoFe₂O₄** for compositions for x = 0.2, 0.4, 0.6 & 0.8. The elements present in the composite are only present, which confirms that the composite is well prepared without impurities. The elemental compositions in terms of mass percentages and atomic percentages are listed in **table 2.11**. The atomic percentages and mass percentages of the prepared samples confirm the stoichiometry of the prepared samples.

3.5.4   (1-x)BaTiO₃ + xMgFe₂O₄

Figure 2.16 shows the energy dispersive X-ray spectroscopy (EDS) spectra of the composite **(1-x)BaTiO₃ + xMgFe₂O₄** for compositions for x = 0.2, 0.4. The elements present in the composite are only present, which confirms that the composite is well prepared without impurities. The elemental compositions in terms of mass percentages and atomic percentages are listed in **table 2.12**. The atomic percentages and mass percentages of the prepared samples confirm the stoichiometry of the prepared samples.

3.6   Optical studies: UV-Vis absorption Spectroscopy

The optical band gap of all the prepared composites of all the compositions have been evaluated using the absorption data obtained using UV-Vis spectrophotometer. From the UV-Vis absorption spectra in the range of 200 nm to 2000 nm, the optical band gap has been evaluated through Tauc plot [Wood and Tauc, 1972] methodology which is explained in **chapter 1, section 1.10.3**.

3.6.1   (1-x)BaTiO₃ + xNiFe₂O₄

The UV-Vis spectrum for all the compositions x = 0.2, 0.4, 0.6 and 0.8 of the composite (1-x)BaTiO₃ + xNiFe₂O₄ is shown in **figure 2.17**. The Tauc plot of all the spectrum are plotted for the extraction of the bandgap of the compositions of the composite. The Tauc plot is shown in **figure 2.18**. The values of the bandgap are tabulated in **table2.13**. The optical bandgap values decrease with the increase in the ferrite and it varies from 1.393 to 1.612 eV.

3.6.2   (1-x)BaTiO₃ + xZnFe₂O₄

The UV-Vis spectrum for all the compositions x = 0.2, 0.4, 0.6 and 0.8 of the composite $(1-x)BaTiO_3 + xZnFe_2O_4$ is shown in **figure 2.19**. The Tauc plot of all the spectrum are plotted for the extraction of the bandgap of the compositions of the composite. The Tauc plot is shown in **figure 2.20**. The values of the bandgap are tabulated in **table2.14**. The optical bandgap values varies from 1.95 eV to 2.68 eV for the compositions from x = 0.2 to 0.8. For x = 0.8, the optical band gap value is maximum.

3.6.3   (1-x)BaTiO₃ + xCoFe₂O₄

The UV-Vis spectrum for all the compositions x = 0.2, 0.4, 0.6 and 0.8 of the composite $(1-x)BaTiO_3 + xCoFe_2O_4$ is shown in **figure 2.21**. The Tauc plot of all the spectrum are plotted for the extraction of the bandgap of the compositions of the composite. The Tauc plot is shown in **figure 2.22**. The values of the bandgap are tabulated in **table2.15**. The optical bandgap value for x = 0.6 is a maximum and for x = 0.8, the bandgap value is a minimum. The optical bandgap varies from 2.22 eV to 2.94 eV.

3.6.4   (1-x)BaTiO₃ + xMgFe₂O₄

The UV-Vis spectrum for all the compositions x = 0.2, 0.4, 0.6 and 0.8 of the composite $(1-x)BaTiO_3 + xCoFe_2O_4$ is shown in **figure 2.23**. The Tauc plot of all the spectrum are plotted for the extraction of the bandgap of the compositions of the composite. The Tauc plot is shown in **figure 2.24**. The values of the bandgap are tabulated in **table2.16**. The optical bandgap decreases from 2.65 eV to 2.27 eV for x = 0.2 to 0.4.

Table 3.5 *Comparison of the range of the optical bandgaps of the composites*

Composites	Optical bandgap (Range) (eV)
BTNF (x = 0.2, 0.4, 0.6, 0.8)	1.39 - 1.61
BTZF (x = 0.2, 0.4, 0.6, 0.8)	1.95 - 2.68
BTCF (x = 0.2, 0.4, 0.6, 0.8)	2.22 - 2.94
BTMF (x = 0.2, 0.4)	2.27 - 2.65

BTNF: $(1-x)BaTiO_3 + xNiFe_2O_4$; BTZF: $(1-x)BaTiO_3 + xZnFe_2O_4$; BTCF: $(1-x)BaTiO_3 + xCoFe_2O_4$; BTMF: $(1-x)BaTiO_3 + xMgFe_2O_4$

3.7   Electrical studies: Dielectric & PE characterization

The electrical characterization for all the composites is investigated from the dielectric and PE characterization. The dielectric characterization is investigated from the capacitance vs frequency graph. Since the capacitance is directly proportional to the dielectric constant. The P-E hysteresis loop is drawn from the P-E characterization data. The P-E hysteresis loop shifts from lossy capacitance response to resistive capacitance response for all the composites [M. Stewart et al., 1999].

3.7.1 $(1-x)BaTiO_3 + xNiFe_2O_4$

The room temperature capacitance vs. frequency is shown in **figure 2.25(a)**. As the value of the capacitance decreases, the dielectric constant also decreases for all the compositions with an increase in frequency. The overall dielectric behavior is of Maxwell–Wagner type [Wagner,1916 Coops, 1951]. At low frequencies, the charge carriers get accumulated at the grain boundaries resulting in a high capacitance. The capacitance decreases almost exponentially with the increase in frequency. This is due to the fact that the charge carriers in the dielectric medium could not follow the high frequency of the alternating field [Roberto Koferstein et al., 2015]. The capacitance decreases with the increase in the ferrite from x = 0.2 to 0.8. The graph between D (dielectric loss) and the frequency is shown in **figure 2.25(b)**. The dielectric loss increases with the increased ferrite content in the composite.

The responses of PE loop of the $BaTiO_3$: $NiFe_2O_4$ composite shown in **figure 2.26** gets deviated from the ideal ferroelectric loop as it appears in oval shape indicating the response of lossy capacitance due to field discharge by conductive ferrite phase . The maximum polarization from the P-E hysteresis loop increases with the increase in the ferrite content and is maximum for x = 0.6 **(0.412 μC/cm²)** But then the remnant polarization becomes higher than the maximum polarization for x = 0.8. The corresponding values are tabulated in **table 2.17**. Hence the composite changes from a lossy capacitance to resistive capacitance for x = 0.8 [M. Stewart et al., 1999].

3.7.2 $(1-x)BaTiO_3 + xZnFe_2O_4$

The room temperature capacitance vs. frequency is shown in **figure 2.27(a)**. As the value of the capacitance decreases, the dielectric constant also decreases for all the compositions with an increase in frequency. The overall dielectric behavior is of Maxwell–Wagner type [Wagner,1916 Coops, 1951]. At low frequencies, the charge carriers get accumulated at the grain boundaries resulting in a high capacitance. The capacitance decreases exponentially with the increase in frequency. This is due to the fact that the charge carriers in the dielectric medium could not follow the high frequency of the alternating field

[Roberto Koferstein et al., 2015]. The capacitance decreases for x = 0.2 to 0.6 and for x = 0.8 it acquires a maximum value. The graph between D (dielectric loss) and the frequency is shown in **figure 2.27(b).** The dielectric loss increases with the increased ferrite content in the composite. Polarization in the material generally is from the dielectric phase. The responses of PE loop of the present $BaTiO_3$: $ZnFe_2O_4$ composite shown in **figure 2.28** deviated from the ideal ferroelectric loop as it appears in oval shape indicating the response of lossy capacitance due to field discharge by conductive ferrite phase [Khamkongkeo, 2011 Pachari, 2015]. The maximum polarization from the P-E hysteresis loop increases with the increase in the ferrite content. But then the remnant polarization which also increases from x = 0.2 to 0.8, becomes higher than the maximum polarization for x = 0.6 and 0.8. The corresponding values are tabulated in Table 7. Hence the composite changes from a lossy capacitance (x = 0.2, 0.4) to resistive capacitance (x = 0.6, 0.8) [M. Stewart et al., 1999].

3.7.3 $(1-x)BaTiO_3 + xCoFe_2O_4$

The room temperature capacitance vs. frequency is shown in **figure 2.29(a).** As the value of the capacitance decreases, the dielectric constant also decreases for all the compositions with an increase in frequency. The overall dielectric behavior is of Maxwell–Wagner type [Wagner,1916 Coops, 1951]. At low frequencies, the charge carriers get accumulated at the grain boundaries resulting in a high capacitance. The capacitance decreases almost exponentially with the increase in frequency. This is due to the fact that the charge carriers in the dielectric medium could not follow the high frequency of the alternating field [Roberto Koferstein et al., 2015]. The capacitance is a maximum for x = 0.2 and minimum for x = 0.4 and 0.8. The graph between D (dielectric loss) and the frequency is shown in **figure 2.29(b).** The dielectric loss increases with the increased ferrite content in the composite. The responses of PE loop of the $BaTiO_3$: $CoFe_2O_4$ composite shown in **figure 2.30** gets deviated from the ideal ferroelectric loop as it appears in oval shape indicating the response of lossy capacitance due to field discharge by conductive ferrite phase. The maximum polarization from the P-E hysteresis loop increases with the increase in the ferrite content and is maximum for x = 0.8 (**4.32 μC/cm²**) But then the remnant polarization becomes higher than the maximum polarization from x = 0.4 to 0.8. The corresponding values are tabulated in **table 2.19**. Hence the composite changes from a lossy capacitance to resistive capacitance from x = 0.4 [M. Stewart et al., 1999].

3.7.4 $(1-x)BaTiO_3 + xMgFe_2O_4$

Figure 2.31(a) exhibits the graph between frequency and capacitance for x = 0, 0.2 and 0.4 compositions of the composite. As the value of the capacitance decreases, the dielectric constant also decreases for all the compositions with an increase in frequency. The overall dielectric behavior is of Maxwell–Wagner type [Wagner,1916 Coops, 1951]. At low frequencies, the charge carriers get accumulated at the grain boundaries resulting in a high capacitance. The capacitance decreases almost exponentially with the increase in frequency. This is due to the fact that the charge carriers in the dielectric medium could not follow the high frequency of the alternating field [Roberto Koferstein et al., 2015]. The constant value of the capacitor for $x = 0$ is minimum which is for pure $BaTiO_3$. As the ferrite is added, for $x = 0.2$ the capacitance is maximum and then decreases for $x = 0.4$. The various polarization contributions decrease as the frequency increases lead to a decrease in the dielectric properties with increasing frequency. As the ferrite content increases the capacitance decreases. **Figure 2.31(b)** depicts the variation of dielectric loss with frequency in which the dielectric loss increases with the increase in the ferrite content.

Table 3.6 *Comparison of maximum and the remnant polarization of the composites*

Composites	P_{max}	P_r	$-P_r$
	$(\mu C/cm^2)$	$(\mu C/cm^2)$	$(\mu C/cm^2)$
BTNF x = 0.2	0.154	0.035	0.039
x = 0.4	0.393	0.175	0.192
x = 0.6	**0.412**	0	0.393
x = 0.8	0.342	**0.475**	**0.481**
BTZF x = 0.2	0.321	0.0647	0.0713
x = 0.4	0.804	0.489	0.531
x = 0.6	2.92	8.5	8.59
x = 0.8	**3.29**	**12.2**	**12.4**
BTCF x = 0.2	3.21	2.63	2.84
x = 0.4	4.18	4.57	4.8
x = 0.6	4.28	8.94	9.1
x = 0.8	**4.32**	**14.2**	**14.3**

BTNF: $(1-x)BaTiO_3 + xNiFe_2O_4$; BTZF: $(1-x)BaTiO_3 + xZnFe_2O_4$; BTCF: $(1-x)BaTiO_3 + xCoFe_2O_4$

From the above comparison table the maximum values for the maximum polarization and the remnant polarization are highlighted with bold. For all the composites the influence of ferrite on the maximum polarization is observed.

3.8   Magnetic studies: M-H characterization

The magnetic studies of the prepared composites have been carried out from the Vibrating Sample Magnetometer data of the magnetization from the M-H characterization. From the M-H characterization the saturation magnetization, the coercivity and the retentivity are found. The magnetic nature of the material is explained from these observed parameters.

3.8.1 $(1-x)BaTiO_3 + xNiFe_2O_4$

M - H measurements are made to study the magnetic behavior of the prepared composite. Studies by the various researchers show that nickel ferrite possesses a soft magnetic behavior with a very low value of coercivity and with saturation magnetization [Garibashyan et al., 2018]. The magnetic studies of the prepared composite $(1-x) BaTiO_3$ + x $NiFe_2O_4$ (x = 0.2, 0.4, 0.6, 0.8) show high saturation magnetization as well as low coercivity having the characteristic of a soft ferromagnetic material. **Figure 2.32(a)** shows the hysteresis curves for all the composites. **Figure 2.32(b)** represents the expanded form of the hysteresis curves. The saturation magnetization, coercivity, and retentivity are presented in **table 2.20**. The increase of the ferrite content increases the saturation magnetization from 2.423 - 23.282 emu/g and decreases the coercivity from 157.52 - 23.87 G. The increase of the saturation magnetization and decrease of the coercivity with the increase of the ferrite content leads to an increase in the soft magnetic nature of the composite.

3.8.2 $(1-x)BaTiO_3 + xZnFe_2O_4$

The magnetic behavior of the prepared composite is studied by the M - H measurements. Studies by the researchers show that Zinc ferrite possesses a small magnetic behavior and also that the magnetization depends on the size of the particle [Mendonca et al., 2012]. The magnetic studies of the prepared composite $(1-x) BaTiO_3$ + x $ZnFe_2O_4$ (x = 0.2, 0.4, 0.6, 0.8) show low saturation magnetization as well as a low coercivity and retentivity having a small ferromagnetic characterization. **Figure 2.33(a)** shows the hysteresis curves for the composite and **Figure 2.33(b)** shows the expanded form of the hysteresis curves. The saturation magnetization, coercivity, and retentivity are tabulated in **table 21**. The saturation magnetization varies from 0.583 emu/g to 1.359 emu/g. Since the ferrite content is low for x = 0.2, the saturation magnetization, coercivity and retentivity also assume a low value. As the ferrite content is high for $x = 0.8$, the saturation magnetization is maximum (1.359 emu/g).

3.8.3 $(1-x)BaTiO_3 + xCoFe_2O_4$

The magnetic studies of the prepared composite $(1-x)BaTiO_3 + xCoFe_2O_4$ show saturation magnetization as well as coercivity having the characteristic of a ferromagnetic material. The magnetic properties of $BaTiO_3$ and XFe_2O_4 ($X = Co/Ni$) particulate composite are investigated [Tadi et al., 2012 Grigalaitis et al., 2014]. The hysteresis curves for all the composites are shown in **figure 2.34**. The saturation magnetization and coercivity are given in **table 2.22**. The increase of the ferrite content increases the saturation magnetization from 13.182 emu/g to 47.260 emu/g. Coercivity varies from 746.38 G to 957.58 G. Coercivity is high for most of the magnetic dipoles do not line up in the same direction of the field for the hysteresis curve measurements and the presence of ferroelectric influences coercivity of the magnetic property. The high coercivity is a characteristic of a hard magnet.

3.8.4 $(1-x)BaTiO_3 + xMgFe_2O_4$

Magnetic characterization of the prepared composite is studied from the M-H measurements. The magnetic studies of the prepared composite $(1-x)BaTiO_3 + xMgFe_2O_4$ ($x = 0.2, 0.4, 1$) and pure $MgFe_2O_4$ shows saturation magnetization which increases with the ferrite content. **Figure 2.35(a)** shows the hysteresis curves for $x = 0.2$ and 0.4 composite and also for pure $MgFe_2O_4$ ($x = 1$). The saturation magnetization is reached fast with more content of $BaTiO_3$ and the $MgFe_2O_4$ reaches slowly around 5000 Gauss. Hence from the curves, it can be inferred that BTO influences the magnetic characterization of the composite. **Figure 2.35(b)** shows the expanded form of the hysteresis curves. These curves show the coercivity and retentivity of the M-H hysteresis. The saturation magnetization, coercivity, retentivity are presented in **table 2.23**. The retentivity and saturation magnetization increases with the ferrite content and is a maximum for pure ferrite. Coercivity assumes a maximum value of 185.56 G for $x = 0.4$ and a minimum value of 92.764 G for pure $MgFe_2O_4$. Hence $MgFe_2O_4$ has a soft magnetic nature when combined with $BaTiO_3$ increases coercivity and develops hard magnetic nature.

Table 3.7 *Comparison of the magnetic parameters of the prepared composites*

Composites	Magnetic parameters of the composites		
	Coercivity H_{ci} (G)	Magnetization M_s (emu/g)	Retentivity M_r x 10^{-3} (emu)
BTNF	51.3 - 157.5 (0.6/0.2)	2.4 - 23.3 (0.2/0.8)	16.2 - 41.1 (0.8/0.4)
BTZF	35.3 - 207.3 (0.2/0.6)	0.58 - 1.4 (0.2/0.8)	0.17 - 5.6 (0.2/0.4)
BTCF	746.4 - 957.8 (0.2/0.8)	13.2 - 47.3 (0.2/0.8)	6.5 - 20.3 (0.2/0.8)
BTMF	93.1 - 185.6 (0.2/0.4)	1.5 - 16.5 (0.2/1.0)	7.2 - 136 (0.2/1.0)

From the above table it is observed that the saturation magnetization increases with the ferrite content. The influence of the ferroelectric is seen in the values of coercivity/retentivity. The minimum and the maximum of the magnetic parameter is shown for all the composites in **table 3.7**. The coercivity and the retentivity is minimum for the composites BTZF, BTCF and BTMF for x = 0.2. The retentivity is maximum for the increased ferrite content for BTCF and BTMF. In BTNF and BTZF composites the retentivity is maximum for x = 0.4.

3.9　Charge density analysis of the magneto-electric composites

The charge density distribution in the unit cell of the prepared composite has been determined by the maximum entropy method (MEM) [Collins, 1982]. The MEM is an exact tool to study the electron density distribution with enhanced resolution. The bonding nature and the distribution of electrons in the bonding region can be visualized using this technique. [Saravanan, 2008] A software package "Practice of Iterative MEM Analysis" (PRIMA) [Momma, 2008] was used for quantitative enumeration of the MEM charge density. The structure factors extracted from the Rietveld refinement [Rietveld, 1969] technique were used in the software PRIMA [Momma, 2008] by considering 64×64×64 pixels along the a, b and c axes of the unit cell. The electron density distribution in the unit cell is constructed through this software separately for barium titanate and nickel ferrite for all the x values (x = 0.2, 0.4, 0.6, 0.8). Mapping of the charge density distribution has been done by applying the visualization software "Visualization for the Electronic and Structural package" (VESTA) [Izumi et al., 2002].

3.9.1 $(1-x)BaTiO_3 + xNiFe_2O_4$

3.9.1.1 Charge density analysis of $BaTiO_3$

Figure 2.36 shows the three-dimensional electron density distribution of $BaTiO_3$ gives the structural view (Tetrahedral) of the unit cell with an isosurface level of 0.45 e/Å^3 in two different orientations. *The isosurface seen around the atoms represents the electron density around the atoms.* **Figure 2.37** exhibits the two-dimensional representation of the electron density in the plane (1 0 0) depicts the bonding between Ba and O atoms of $BaTiO_3$. Similarly, the bonding between Ti and O is shown in Fig.5 for (2 0 0) plane. The two-dimensional electron density distribution is drawn with contour lines varying from 0 – 1 e/Å^3 with an interval of 0.04 e/Å^3 using the software VESTA [Izumi et al., 2002]. *The density and the shape of the contours show the electron density distribution and the interaction of the atoms.* It can be observed from **figure 2.37** that the contour lines fade amidst 'Ba' and 'O' atoms and the shape of the contour lines around the atoms change with the increase of the ferrite content, indicating an interaction between the atoms. *The low density of the contour lines between the atoms depicts that the bonding is more ionic.*

Figure 2.38 shows the bonding between 'Ti' and 'O' for all the compositions. *The density of the contour lines between the atoms is high, indicating a more covalent nature of the bonding.* Figure 2.39 exhibits the one-dimensional electron density profile which gives the quantitative values of electron density for all the compositions. The values of the electron densities for Ba-O and Ti-O bonds are tabulated in table 2.24. The electron density at the bond critical point for all compositions of the Ti-O bond lies in the range 0.41 to 1.04 e/Å^3, which is quite higher than the electron density of the Ba-O bond ($0.27 - 0.32$ e/Å^3) of the composite. This indicates an increase in the covalence of the bonding between Ti-O.

3.9.1.2 Charge density analysis of $NiFe_2O_4$

Three-dimensional electron density distribution of $NiFe_2O_4$ that depicts the structural view (cubic) of the unit cell with an isosurface level of 0.7e/$\text{Å}3$ and 1.7e/$\text{Å}3$ is shown in **figure 2.40.** The two-dimensional electron density distribution is drawn with contour lines varying from 0 – 0.8 e/Å^3 with an interval of 0.04 e/Å^3 using the software VESTA [Izumi et al., 2002]. **Figure 2.41** shows the two-dimensional representation of the electron density in the plane (1 0 1) depicts the bonding of Ni-O and Fe-O of $NiFe_2O_4$. The density and the shape of the contours show the electron density distribution and the interaction of the atoms. Low-density contour lines between the atoms depict the bonding to be more ionic.

The bonding between 'Ni' and 'O' for all the compositions shows the interaction between the atoms and the density of the contour lines between the atoms being high indicate the nature of the bonding to be more covalent. **Figure 2.42** exhibits the one-dimensional electron density profiles for all the compositions along Fe-O bond and Ni-O bond. The values of the electron density at the bond critical point for Fe-O, and Ni-O are given in **table 2.25**. The electron density at the bond critical point for Fe-O bond for all the compositions vary from 0.16 e/$\AA^3$ to 0.27 e/$\AA^3$ is less compared to the values of Ni-O bond (1.88 to 2.59 e/$\AA^3$), indicating that the bond is more ionic. The electron density at the bond critical point of Ni-O bond, shows a more covalent nature of the bonding. The covalent bonding between the atoms is responsible for the dielectric polarization of the composite. So, the presence of ferrites enhances the dielectric nature of the composite.

3.9.2 $(1-x)BaTiO_3 + xZnFe_2O_4$

3.9.2.1 Charge density studies of $BaTiO_3$

The three-dimensional electron density distribution of $BaTiO_3$ is tetragonal with P4mm space group of the unit cell with an isosurface level of 0.9 e/$\AA^3$ is shown in **figure 2.43** in two different orientations. The isosurface around the atoms represents the electron density around the atoms. **Figure 2.44** exhibits the two-dimensional representation of the electron density in the plane (1 0 0) which depicts the bonding between Ba and O atoms of $BaTiO_3$. **Figure 2.45** shows the bonding between Ti and O in the (2 0 0) plane. The two-dimensional electron density distribution is mapped with contour lines varying from 0 – 0.8 e/$\AA^3$ with an interval of 0.04 e/$\AA^3$ using software VESTA [Izumi et al., 2002]. The density and the shape of the contours show the electron density distribution and the interaction of the atoms. It is observed from **figure 2.44** that the density of contour lines between 'Ba' and 'O' atoms is low and the shape of the contour lines around the atoms change with the increase of the ferrite content, indicating an interaction between the atoms. The low density of the contour lines between the atoms reveals that the bonding is more ionic. For ionic bonds, the ideal electron density is zero at the bond critical point. But theoretically due to the influence of other atoms, minimum electron density is observed.

The density of the contour lines between the atoms is high as the overlapping region between the atoms is seen, indicating a more covalent nature of the bonding. For x = 0.8, the electron density at the bond critical point is minimum indicating that the bonding is more ionic. The one-dimensional electron density profile which gives the quantitative analysis of the electron density is shown in **figure 2.46**. The corresponding values are tabulated in **table 2.26**. The electron density at the bond critical point for all compositions of the Ti-O bond varies from 0.482 - 1.235 e/$\AA^3$, which is higher than the electron density of the Ba – O bond (0.145 - 0.385 e/$\AA^3$) of the composite. This indicates an increase in the covalence of the bonding between Ti -O. If the electron density at the bond critical point is less, it is more ionic; similarly if the electron density is high at the bond critical point, the bonding is more covalent. Hence the Ti-O bond is more covalent.

3.9.2.2 Charge density studies of $ZnFe_2O_4$

The three-dimensional electron density distribution of $ZnFe_2O_4$ with Fd-3m space group, cubic structure of the unit cell with an isosurface level of 1.8 e/$Å^3$ is shown in **figure2.47**. The polygons associated with Zn and Fe which are tetrahedral and octahedral are depicted in **figure 2.47(a)**. The two-dimensional electron density distribution is drawn with contour lines varying from 0 – 0.8 e/$Å^3$ with an interval of 0.04 e/$Å^3$ using software VESTA [Izumi et al., 2002]. **Figure 2.48** shows the two-dimensional representation of the electron density in the plane (1 0 1) and it depicts the bonding of Zn-O and Fe-O of $ZnFe_2O_4$. High density contour lines or overlapping of contour lines indicates that the electron density is high. The one-dimensional electron density profiles for all the compositions along Fe-O bond and Zn-O bond is shown in **figure 2.49**. The values of the electron density at the bond critical point for Fe-O, and Zn-O are given in **table 2.27**. For x = 0.4 and 0.6, the electron density at the bond critical point is 0.596 and 0.662 e/$Å^3$ tends towards being more ionic. For x = 0.2 and 0.8, the electron density at the bond critical point of Zn – O bond being high shows a more covalent nature of the bonding. The electron density between Fe and O increases from x = 0.2 to x = 0.6 and it decreases for x = 0.8 and varies from 0.353 to 0.864 e/$Å^3$ which is more ionic in nature.

3.9.3   (1-x)BaTiO₃ + xCoFe₂O₄

3.9.3.1 Charge density studies of BaTiO₃

The three-dimensional electron density distribution of $BaTiO_3$ with (2 0 0) plane gives the structural view of the unit cell with an isosurface level of 0.45 and 0.8 e/$Å^3$ is shown in **figure 2.50**. Electron density in the plane (1 0 0) is a two-dimensional representation that depicts the bonding between Ba and O atoms and Ti-O of $BaTiO_3$ and is shown in **figure 2.51 and figure 2.52**. The two-dimensional electron density distribution is drawn with contour lines varying from 0 – 1 e/$Å^3$ with an interval of 0.08 e/$Å^3$ using the software VESTA [Izumi et al., 2002]. The density and the shape of the contours show the electron density distribution and the interaction of the atoms. It can be observed from **figure 2.51** that the contour lines fade amidst 'Ba' and 'O' atoms and the shape of the contour lines around the atoms change with the increase of the ferrite content indicating an interaction between the atoms. *The low density of the contour lines between the atoms depicts that the bonding to be more ionic.* **Figure 2.52** shows the bonding between 'Ti' and 'O' for all the compositions in which t*he densities of the contour lines between the atoms are high* *indicating the nature of the bonding to be more covalent.* **Figure 2.53** exhibits the one-dimensional electron density profile which gives the quantitative values of electron density for all the compositions and the values are tabulated in **table 2.28**. The electron density at the bond critical point of the Ti-O bond is high indicating the bonding to be more covalent. Similarly, the electron density at the bond critical point for Ba – O bond is low indicating the bond to be more ionic. *The Ti-O bond plays an important role in the ferroelectric property of the composite.*

3.9.3.2 Charge density studies of $CoFe_2O_4$

The three-dimensional electron density distribution of $CoFe_2O_4$ with (1 0 1) plane gives the structural view of the unit cell with an isosurface level of 0.5 and 1.8 $e/Å^3$ is shown in **figure 2.54**. The two-dimensional representation of electron density in the plane (1 0 1) depicts the bonding of Co-O and Fe-O of $CoFe_2O_4$ is shown in **figure 2.55**. The two-dimensional electron density distribution is drawn with contour lines varying from $0 – 1.8$ $e/Å^3$ with an interval of 0.08 $e/Å^3$ using the software VESTA [Izumi et al., 2002]. The density and the shape of the contours show the electron density distribution and the interaction of the atoms and the low density of the contour lines between the atoms depicts that the bonding to be more ionic. **Figure 2.56** exhibits the one-dimensional electron density profiles for all the compositions along Fe-O bond. The bonding between 'Co' and 'O' for all the compositions shows the interaction between the atoms. *As the ferrite phase increases, the density of the contour lines between the atoms is high indicating the nature of the bonding to be more covalent.* The values of the electron density at the bond critical point for Fe-O, and Co-O are tabulated in **table 2.29**.

3.9.4 $(1-x)BaTiO_3 + xMgFe_2O_4$

3.9.4.1 Charge density studies of $BaTiO_3$

The 3D electron density distribution of $BaTiO_3$ which assumes the tetragonal structure; the unit cell with an isosurface level of 0.5 $e/Å^3$ is shown in **figure 2.57**. **Figure 2.58** exhibits the two-dimensional representation of the electron density in the plane (1 0 0) which depicts the bonding between Ba and O atoms of $BaTiO_3$ for x = 0.2, 0.4 and bonding between Ti and O in the (2 0 0) plane is depicted in **figure 2.59** for x = 0.2, 0.4 with the contour lines varying from $0 – 0.8$ $e/Å^3$ with an interval of 0.04 $e/Å^3$ using software VESTA [Izumi et al., 2002. The density of contour lines between 'Ba' and 'O' atoms is low and the shape of the contour lines around the atoms changes with the increase of the ferrite content, due to the interaction between the atoms. Similarly, the density of the contour lines between Ti-O is high and overlapping and the bonding is more covalent. The one-dimensional electron density profile for all the compositions x = 0.0, 0.2 & 0.4 are shown in **figure 2.60**.

The values of the electron density for all the compositions are tabulated in **table 2.30**. At the bond critical point the electron density for the compositions $x = 0.2$ and 0.4 of the Ti-O bond is 0.470 and 0.471 e/$\mathring{A}^3$, which is higher than the electron density of the Ba-O bond (0.202 and 0.154 e/$\mathring{A}^3$) of the $x = 0.2$ and 0.4 of the composite. The charge density at the bond critical point of Ba-O and Ti-O of the prepared pure $BaTiO_3$ are 0.287 and 0.113 e/$\mathring{A}^3$. Hence, the values of the electron density at the bond critical point decreases for Ba-O bond and increases for Ti-O bond with the increase in the ferrite content from $x = 0.2$ to 0.4. The lower electron density at the bond critical point between two atoms represent more ionic and vice versa.

3.9.4.2 Charge density studies of MgFe₂O₄

The 3D electron density distribution of $MgFe_2O_4$ unit cell is shown in **figure 2.61**. The polygons associated with Mg and Fe are shown in the figure, where the Mg assumes tetrahedral polygon and Fe assumes octahedral polygon. The 2D electron density distribution is drawn with contour lines varying from $0 - 0.8$ e/$\mathring{A}^3$ with an interval of 0.04 e/$\mathring{A}^3$ using the software VESTA [Izumi et al., 2002]. **Figure 2.62** shows the two-dimensional representation of the electron density in the plane ($1\ 0\ 1$) which depicts the bonding of Mg-O and Fe-O of $MgFe_2O_4$ for $x = 0.2$ and 0.4. One-dimensional electron density profiles for all the compositions $x = 0.2$, 0.4 and 1.0 along Fe-O and Mg-O bond are shown in **figure 2.63**. The values of the electron density at the bond critical point for Fe-O and Mg-O are given in **table2.31**.

The values of the electron density at the bond critical point along Fe-O, and Mg-O bonds of pure $MgFe_2O_4$ are 0.165 and 0.599 e/$\mathring{A}^3$. The electron density at the bond critical point along Fe-O bond for $x = 0.2$ and 0.4 are 0.528 and 0.337 e/$\mathring{A}^3$ which decreases with the increase in the ferrite content and for Mg-O is 0.488 and 0.706 e/$\mathring{A}^3$ for $x = 0.2$ and 0.4. It can be inferred that the electron density at the bond critical point of Mg-O bond, increases with the increase in the ferrite content. *The electron density of the Mg-O bond of pure MgFe₂O₄ is low compared to the bond for x = 0.4, hence it can be inferred that the presence of BaTiO₃ does influence the electron density at the bond critical point of MgFe₂O₄.*

Table 3.8 *Comparison of the charge densities of the prepared composites*

Composites	Charge densities at the bond critical point (e/Å3)			
	Ba-O (xMin/Max)	Ti-O (xMin/Max)	Fe-O (xMin/Max)	M-O (xMin/Max)
BTNF	0.28 - 0.32 (0.6/0.4)	0.42 - 1.05 (0.4/0.8)	0.16 - 0.27 (0.4/0.6)	1.82 - 2.59 (0.8/0.6)
BTZF	0.15 - 0.39 (0.2/0.8)	0.48 - 1.41 (0.8/0.2)	0.35 - 0.86 (0.8/0.6)	0.60 - 2.8 (0.4/0.8)
BTCF	0.15 - 0.25 (0.6/0.2)	1.08 - 1.46 (0.2/0.4)	0.24 - 0.49 (0.8/0.6)	0.72 - 2.39 (0.6/0.8)
BTMF	0.154 - 0.287 (0.4/0)	0.113 - 0.471 (0/0.4)	0.17 - 0.53 (1.0/0.2)	0.49 - 0.71 (0.2/0.4)

M: Ni/Zn/Co/Mg ; BT: $BaTiO_3$ NF: $NiFe_2O_4$ ZF: $ZnFe_2O_4$ CF: $CoFe_2O_4$ MF: $MgFe_2O_4$

From the above table it can be clearly observed that the Ba-O and Fe-O bonds are more ionic, Ti-O and M-O bonds are more covalent. In a composite, all the four bonds play a role in the properties of the material. From the observations made, the high electron density leads to low coercivity of the material. Coercivity is the magnetic property of the material which interprets the soft / hard magnetic nature of the material.

References

• Collins D. M, Nature. 298, 49 (1982). https://doi.org/10.1038/298049a0

• Garibashiyan M., N. Sadegh, and N. Omid Mirzae, J. SolGel Sci. Technol. 85, 1 (2018).

• Grigalaitis.R, Vijatović Petrović M.M., Bobićb J.D., Dzunuzovic A., Sobiestianskas R., Brilingas A.,.Stojanović B.D, Banysa J., Ceramics International 40, 6165 (2014) https://doi.org/10.1016/j.ceramint.2013.11.069

• Izumi F, Dilanien R. A, Recent Research Developments in Physics Part II, Vol.3, Transworld Research Network. Trivandrum, 699-726, (2002).

• Khamkongkaeo A., Jantaratana P., Sirisathitkul C., Yamwong T., Maensiri S., Frequency-dependent magnetoelectricity of CoFe2O4-BaTiO3 particulate composites, Trans. Nonferrous Met. Soc. China 21 (11) 2438(2011). https://doi.org/10.1016/S1003-6326(11)61033-9

• Koops C., On the dispersion of resistivity and dielectric constant of some semiconductors at audiofrequencies, Phys. Rev. 83 (1) (1951) 121-124. https://doi.org/10.1103/PhysRev.83.121

• Mendonça E.C., Jesus C.B.R., Folly W.S.D., Meneses C.T., Duque J.G.S., Size effects on the magnetic properties of ZnFe2O4 nanoparticles, J. Supercond. Nov. Magn. 26(6), 2329 (2013). https://doi.org/10.1007/s10948-012-1426-3

• Momma K, Izumi F, VESTA: a three-dimensional visualization system for electronic and structural analysis. J. Appl. Crystallogr. 41, 653 (2008). https://doi.org/10.1107/S0021889808012016

• Pachari S., Structure, Microstructure and Magneto-Dielectric Properties of Barium Titanate-ferrite Based Composites, National Institute of Technology Rourkela, Rourkela, (2015)

• Petricek V, Dusek M, Palatinus L, Kristallogr Z, Crystallographic Computing System JANA 2006: General features, 229, 345 (2014). https://doi.org/10.1515/zkri-2014-1737

• Roberto Köferstein, Till Walther, Dietrich Hesse, Stefan Ebbinghaus, Fine-grained BaTiO3 + MgFe2O4 composites prepared by a Pechini-like process, J. Alloy. Compd. 638 141(2015) https://doi.org/10.1016/j.jallcom.2015.03.082

• Saravanan R, GRAIN software, Personal communication (2008)

• Saravanan R., Practical application of maximum entropy method in electron density and bonding studies, Phys. Scr. 79 048303 (2009) https://doi.org/10.1088/0031-8949/79/04/048303

• Stevart M., Cain M G., Hall D A (1999) https://physlab.lums.edu.pk/images/e/eb/Reframay4.pdf

• Tadi, R., Kim, Y.I., Ryu, K.S. & Kim, C. Journal of the Korean Physical Society 61(9) 1545 (2012) https://doi.org/10.3938/jkps.61.1545

• Wagner K.W., Zur theorie der unvollkommenen dielektrika, Ann. Phys. 345 (5) 817(1913). https://doi.org/10.1002/andp.19133450502

• Wood D. L, Tauc J, Phys Rev B. 5, 3144 (1972). https://doi.org/10.1103/PhysRevB.5.3144

• Wyckoff R.W.G, Crystal structures, Vol.1, Inter-space publishers, London (1963).

Chapter 4

Conclusion

Abstract

In this work, four magneto-electric ceramic composite series of ferroelectric and ferrite materials, I. (1-x) BaTiO$_3$ + x NiFe$_2$O$_4$, II (1-x) BaTiO$_3$ + x ZnFe$_2$O$_4$, III (1-x) BaTiO$_3$ + x CoFe$_2$O$_4$,IV(1-x) BaTiO$_3$ + x MgFe$_2$O$_4$ were synthesized. The structural parameters of the prepared magneto-electric ceramic composites are investigated from the powder X-ray diffraction data recorded over the range of 10° to 120°. From the morphological studies, the average and the individual particle sizes of the composite are found. The optical characterization of the prepared composites for all the compositions is analyzed from the data UV-Vis spectrum. The dielectric characterization reveals the dependence of the capacitance on the frequency and on the composition of the ferrite. The dielectric loss increases with the increase in the ferrite.The M-H characterization of the prepared composite for all the compositions is observed for all the compositions. The saturation magnetization increases with the ferrite content and the influence of the ferroelectric is observed in the coercivity. The electron density distribution in the three-dimensional view has been analyzed and it gives the structure of barium titanate and the ferrite.

Conclusion

In this work, four magneto-electric ceramic composite series of ferroelectric and ferrite materials,

I (1-x) BaTiO$_3$ + x NiFe$_2$O$_4$ **II (1-x) BaTiO$_3$ + x ZnFe$_2$O$_4$**

III (1-x) BaTiO$_3$ + x CoFe$_2$O$_4$ **IV(1-x) BaTiO$_3$ + x MgFe$_2$O$_4$**

were synthesized and investigated using powder X-ray diffraction (PXRD), scanning electron microscopy (SEM) and energy dispersive X-ray spectroscopy (EDS), UV-visible spectrophotometry (UV-Vis), Electrical (Dielectric and P-E) characterization, Magnetic characterization (M-H). The results obtained from the different experimental characterization and analytical techniques are analyzed and summarized.

(i) Structural

The structural parameters of the prepared magneto-electric ceramic composites are investigated from the powder X-ray diffraction data recorded over the range of 10° to 120°. The structure of the ferroelectric (BaTiO$_3$) is tetragonal with P4mm space group and the structure of the ferrites are cubic with space group Fd-3m. Structure of the $BaTiO_3$ is influenced with the presence of ferrite. In $BaTiO_3$: $NiFe_2O_4$ composite, the lattice parameters of $BaTiO_3$ for x = 0.4 composition is maximum and the c/a ratio is high for x = 0.8. In $BaTiO_3$: $ZnFe_2O_4$ composite, the lattice parameters of $BaTiO_3$ for x = 0.8 composition is maximum and the c/a ratio is high for x = 0.4. In $BaTiO_3$: $CoFe_2O_4$ composite, the lattice parameters of $BaTiO_3$ and the c/a ratio is maximum for x = 0.2. In $BaTiO_3$: $MgFe_2O_4$ composite, the lattice parameters of $BaTiO_3$ and the c/a ratio is maximum for pure $BaTiO_3$ (x = 0). For $BaTiO_3$: $MgFe_2O_4$ is analyzed for x = 0, 0.2, 0.4 and 1.0. The structure of the ferrites assuming cubic shape is also influenced by the presence of ferroelectric material. The lattice parameter of $NiFe_2O_4$ is maximum for x = 0.6, for $ZnFe_2O_4$ and $CoFe_2O_4$, the lattice parameter is maximum for x = 0.2. For $MgFe_2O_4$ the lattice parameter is maximum for x = 1.

(ii) Morphological Studies

From the morphological studies, the average and the individual particle sizes of the composite are found. The average particle size of the composites for the compositions x = 0.2, 0.4, 0.6 and 0.8 varies from 0.97 to 3.4 µm. The individual particle size of $BaTiO_3$ varies from 17.46 to 34.98 nm and for the ferrites, the individual particle size varies in the range 16.24 to 39.42 nm.

(iii) Elementary analysis

The elementary analysis from the energy dispersive X-ray spectrum of all the composites confirms the presence of only the elements in the materials used for the preparation of the composite and there are no impurities present in the ceramic composite

(iv) Optical Studies

The optical characterization of the prepared composites for all the compositions is analyzed from the data UV-Vis spectrum. The optical bandgap of BTNF takes a lower value and a lower range for the composites and a maximum value for BTCF. The lower band gap value indicates that the material is more a semiconductor having the probability of acquiring the property of conductivity.

(v) Electrical Studies

The dielectric characterization reveals the dependence of the capacitance on the frequency and on the composition of the ferrite. The dielectric loss increases with the increase in the ferrite. The P-E hysteresis loop shifts from lossy capacitance to a resistive capacitance response for all the composites, The maximum polarization is influenced by the presence of the ferrite, it increases with the increase in the ferrite content. For BTNF and BTZF the shift from the lossy capacitance to resistive capacitance is observed for x = 0.6 and 0.8. The ferrite influences the characteristic of the ferroelectric material.

(vi) Magnetic Studies

The M-H characterization of the prepared composite for all the compositions is observed for all the compositions. The saturation magnetization increases with the ferrite content and the influence of the ferroelectric is observed in the coercivity. The coercivity and the retentivity are minimum for the composites BTZF, BTCF and BTMF for x = 0.2. The retentivity is maximum for the increased ferrite content for BTCF and BTMF. In BTNF and BTZF composites the retentivity is maximum for x = 0.4.

(vii) Charge density analysis

The bonding nature and the distribution of electrons in the bonding region are analyzed for Ba-O, Ti-O bonds of $BaTiO_3$ and Fe-O, M-O bonds in MFe_2O_4 (**M** = Ni, Zn, Co, Mg) which form the composite of different series. The electron density distribution in the three-dimensional view has been analyzed and it gives the structure of barium titanate and the ferrite. The qualitative analysis of the charge density is investigated from the two-dimensional electron density distributions between the atoms. The bonding between Ba-O and Fe-O are found to be more ionic. The bonding between Ti-O and M-O are more covalent. The corresponding values of the charge density at the bond critical point are elucidated from the one-dimensional charge density profile. The quantitative values and the qualitative results agree with each other.

About the Author

Dr Ramachandran Saravanan, has been associated with the Department of Physics, The Madura College, affiliated with the Madurai Kamaraj University, Madurai, Tamil Nadu, India from the year 2000. He is the head of the Research Centre and PG department of Physics. He worked as a research associate during 1998 at the Institute of Materials Research, Tohoku University, Sendai, Japan and then as a visiting researcher at Centre for Interdisciplinary Research, Tohoku University, Sendai, Japan up to 2000.

Earlier, he was awarded the Senior Research Fellowship by CSIR, New Delhi, India, during Mar. 1991 - Feb.1993; awarded Research Associateship by CSIR, New Delhi, during 1994 – 1997. Then, he was awarded a Research Associateship again by CSIR, New Delhi, during 1997- 1998. Later he was awarded the Matsumae International Foundation Fellowship in1998 (Japan) for doing research at a Japanese Research Institute (not availed by him due to the simultaneous occurrence of other Japanese employment).

He has guided twelve Ph.D. scholars as of 2018, and about six researchers are working under his guidance on various research topics in materials science, crystallography and condensed matter physics. He has published around 150 research articles in reputed Journals, mostly International, apart from around 50 presentations in conferences, seminars and symposia. He has also guided around 60 M.Phil. scholars and an equal number of PG students for their projects. He has attracted government funding in India, in the form of Research Projects. He has completed two CSIR (Council of Scientific and Industrial Research, Govt. of India), one UGC (University Grants Commission, India) and one DRDO (Defense Research and Development Organization, India) research projects successfully and is proposing various projects to Government funding agencies like CSIR, UGC and DST.

He has written 12 books in the form of research monographs including; "Experimental Charge Density - Semiconductors, oxides and fluorides" (ISBN-13: 978-3-8383-8816-8; ISBN-10:3-8383-8816-X), "Experimental Charge Density - Dilute Magnetic Semiconducting (DMS) materials" (ISBN-13: 978-3-8383-9666-8; ISBN-10: 3-8383-9666-9) and "Metal and Alloy Bonding - An Experimental Analysis" (ISBN -13: 978-1-4471-2203-6). He has committed to write several books in the near future.

His expertise includes various experimental activities in crystal growth, materials science, crystallographic, condensed matter physics techniques and tools as in slow evaporation, gel, high temperature melt growth, Bridgman methods, CZ Growth, high vacuum sealing etc. He and his group are familiar with various equipment such as: different types of cameras; Laue, oscillation, powder, precession cameras; Manual 4-circle X-ray diffractometer, Rigaku 4-circle automatic single crystal diffractometer, AFC-5R and AFC-

7R automatic single crystal diffractometers, CAD-4 automatic single crystal diffractometer, crystal pulling instruments, and other crystallographic, material science related instruments. He and his group have sound computational capabilities on different types of computers such as: IBM – PC, Cyber180/830A – Mainframe, SX-4 Supercomputing system – Mainframe. He is familiar with various kind of software related to crystallography and materials science. He has written many computer software programs himself as well. Around twenty of his programs (both DOS and GUI versions) have been included in the SINCRIS software database of the International Union of Crystallography.